ROUGH-HEWN LAND

THE PUBLISHER GRATEFULLY ACKNOWLEDGES THE GENEROUS
SUPPORT OF THE AUGUST AND SUSAN FRUGÉ ENDOWMENT FUND
IN CALIFORNIA NATURAL HISTORY OF THE UNIVERSITY OF
CALIFORNIA PRESS FOUNDATION.

ROUGH-HEWN LAND

*A Geologic Journey from California
to the Rocky Mountains*

Keith Heyer Meldahl

UNIVERSITY OF CALIFORNIA PRESS

Berkeley   Los Angeles   London

University of California Press, one of the most distinguished university presses in the United States, enriches lives around the world by advancing scholarship in the humanities, social sciences, and natural sciences. Its activities are supported by the UC Press Foundation and by philanthropic contributions from individuals and institutions. For more information, visit www.ucpress.edu.

For a digital version of this book, see the press website.

University of California Press
Berkeley and Los Angeles, California

University of California Press, Ltd.
London, England

Library of Congress Cataloging-in-Publication Data

Meldahl, Keith Heyer.
  Rough-hewn land : a geologic journey from California to the Rocky Mountains / Keith Heyer Meldahl.
      p. cm.
  Includes bibliographical references and index.
  ISBN 978-0-520-25935-5 (hardback)
  1. Geology—West (U.S.) 2. West (U.S.)—History. I. Title.
  QE79.M45 2011
  557.9—dc23
                                    2011026058

19   18   17   16   15   14   13   12   11
10   9   8   7   6   5   4   3   2   1

Cover image: View across Walker Lake to Mount Grant, Wassuk Range, Great Basin, Nevada. Photo © Scott T. Smith.

For Malcolm, Jim, Dan, and Jim

May your trails be dim, lonesome, stony, narrow, winding, and only slightly uphill.
May God's dog serenade your campfire, and may the rattlesnake
and the screech owl amuse your reverie.
May the Great Sun dazzle your eyes by day
and the Great Bear watch over you by night.

EDWARD ABBEY, 1971

Science is nothing but trained and organized common sense, differing from the latter only as a veteran may differ from a raw recruit: and its methods differ from those of common sense only as far as the guardsman's cut and thrust differ from the manner in which a savage wields his club.

THOMAS H. HUXLEY, DARWIN DEFENDER AND EVOLUTION PROMOTER,
ON THE NATURE OF THE SCIENTIFIC METHOD, 1893

We are like a judge confronted by a defendant who declines to answer, and we must determine the truth from the circumstantial evidence.

ALFRED WEGENER, DISCOVERER OF CONTINENTAL DRIFT,
ON THE CHALLENGE OF INTERPRETING THE EARTH'S PAST
FROM EVIDENCE IN ROCKS, 1929

Geology is not a totally exact science; you have to be creative. For the most part, we don't have enough information to know everything we'd like to know about what's going on. So you have to fill in the gaps, think outside the box, release your inhibitions—and beer is one way to do that.

RICK SALTUS, U.S. GEOLOGICAL SURVEY,
ON WHY GEOLOGISTS DRINK SO MUCH BEER, 2009

CONTENTS

Preface · xv

PART I: CALIFORNIA · 1

1. Golden Gate · 3

2. Mother Lode · 19

3. Rivers of Gold · 35

4. A Traverse across the Range of Light · 47

5. Where Is the Edge of the North American Plate? · 61

PART II: THE BASIN AND RANGE AND THE GREAT BASIN · 77

6. Where Rivers Die · 79

7. The Growing Pains of Mountains · 97

8. Wealth and Magma · 117

9. Water and Salt · 131

10. Evolution's Big Bang · 147

PART III: THE ROCKY MOUNTAINS · 163

11. Range-Roving Rivers · 165

12. Up from the Basement · 183

13. At the Frontier · 199

Appendix I: Deep Time: Fathoming the Rock Record · 217
Appendix II: Seeing for Yourself · 233
Acknowledgments · 245
Notes · 247
Glossary · 259
Bibliography · 267
Figure Sources and Credits · 281
Index · 287

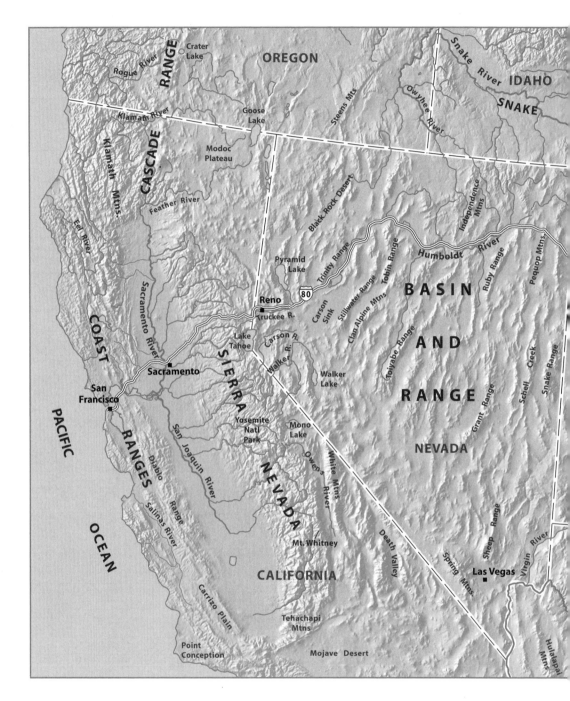

This book traverses a broad swath across the central western United States centered roughly on Interstate Highway 80. We begin at the Golden Gate by San Francisco, pass through California's Coast Ranges and Central Valley, ascend the Sierra Nevada, drop into the immense bowl of the Great Basin (part of the Basin and Range Province), climb the Rocky Mountains, and end on the western Great Plains.

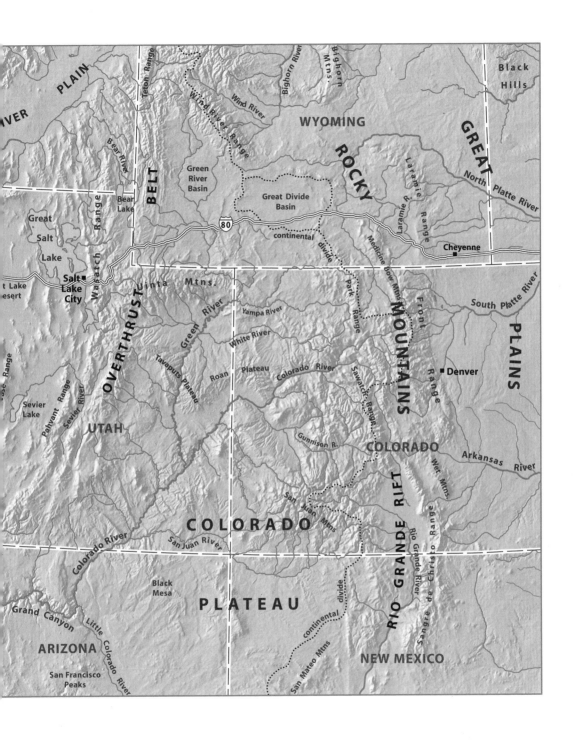

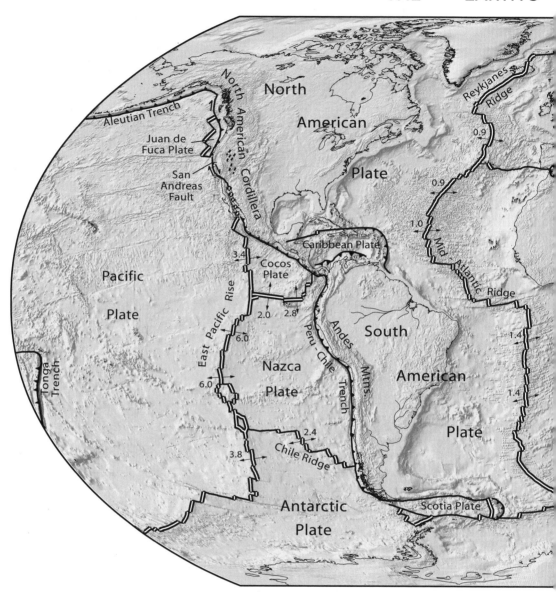

North American Plate

North American Cordillera

Aleutian Trench

Juan de Fuca Plate

1

San Andreas Fault

Reykjanes Ridge

0.9

0.9

1.0

Mid-Atlantic Ridge

Caribbean Plate

Cocos Plate

3.4

2.0   2.8

Pacific

Plate

East Pacific Rise

6.0

6.0

Nazca Plate

Peru-Chile Trench

Andes Mtns

South American Plate

1.4

1.4

Tonga Trench

2.4

3.8   Chile Ridge

Antarctic Plate

Scotia Plate

Mid-ocean ridges (where the seafloor spreads) and transform faults (faults that link offset segments of ridges)

2.4   Total spreading rate in inches per year

Major continental rift zones

TECTONIC PLATES

North American Plate

Eurasian   Plate

Ural Mtns.

Altyn Tagh fault

Zagros Mtns.   Himalaya

Arabian
Plate   Pacific

Philippine   Plate
Plate

African   Indian   Mariana Trench

1.1

Caroline
Plate

Somalia   1.2   Plate
Plate

East African Rift Valley

Plate

1.8   Java Trench

0.6

Australian   Plate

S.W. Indian
Ocean Ridge   S.E. Indian Ocean Ridge

0.6   2.9

3.0

Antarctic   Plate

⇄ Major transform faults (side-by-side moving faults) with
arrows showing sense of motion

▬ᴧᴧᴧ Ocean trenches and continental collision zones; barbs show
the direction that one plate is plunging under the other

⌄⌄? Plate boundary uncertain

EON	ERA	PERIOD	EPOCH	AGE	CALIFORNIA: COASTAL REGION
Phanerozoic Eon	Cenozoic	Quaternary	Holocene + Pleistocene	2.6	San Andreas fault system (still active)
		Neogene	Pliocene	5.3	
			Miocene	23	
		Paleogene	Oligocene	34	
			Eocene	56	accretion of terranes of the Franciscan Complex (deep sea rock scraped off the Farallon Plate)
			Paleocene	66	
	Mesozoic	Cretaceous		146	
		Jurassic		202	
		Triassic		251	
	Paleozoic	Permian		299	
		Pennsylvanian		318	
		Mississippian		359	region did
		Devonian		416	
		Silurian		444	
		Ordovician		488	
		Cambrian		542	
Proterozoic Eon				2500	
Archean Eon				3850	
Hadean Eon				4550	

Major geologic developments in the American West from the California coast to the Rocky Mountains. Geologic ages are in millions of years. Note that the vertical scale varies; older geologic intervals cover greater time spans than do younger intervals.

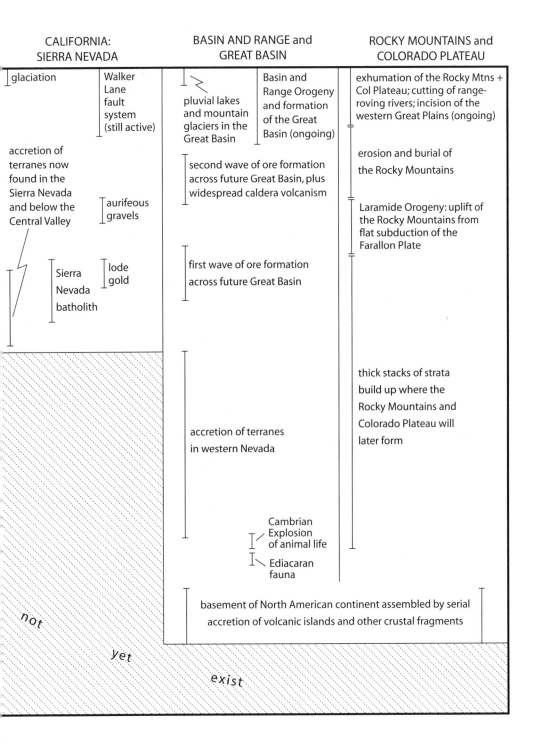

CALIFORNIA: SIERRA NEVADA		BASIN AND RANGE and GREAT BASIN		ROCKY MOUNTAINS and COLORADO PLATEAU
glaciation	Walker Lane fault system (still active)	pluvial lakes and mountain glaciers in the Great Basin	Basin and Range Orogeny and formation of the Great Basin (ongoing)	exhumation of the Rocky Mtns + Col Plateau; cutting of range-roving rivers; incision of the western Great Plains (ongoing)
accretion of terranes now found in the Sierra Nevada and below the Central Valley	aurifeous gravels	second wave of ore formation across future Great Basin, plus widespread caldera volcanism		erosion and burial of the Rocky Mountains
	lode gold	first wave of ore formation across future Great Basin		Laramide Orogeny: uplift of the Rocky Mountains from flat subduction of the Farallon Plate
Sierra Nevada batholith				
		accretion of terranes in western Nevada		thick stacks of strata build up where the Rocky Mountains and Colorado Plateau will later form
			Cambrian Explosion of animal life	
			Ediacaran fauna	
		basement of North American continent assembled by serial accretion of volcanic islands and other crustal fragments		

not yet exist

PREFACE

The mountains are calling me and I must go.

JOHN MUIR, 1873

Twenty-seven years ago, I climbed into my '67 Chevy on the east coast and aimed it toward the Pacific Ocean. Everything I owned fit in the trunk with room to spare for twelve quarts of motor oil that would be gone before I reached California. I had a fresh geology degree tucked between my ears, but when it came to the American West, I was as ignorant as a stone. The world west of the Rockies was *terra incognita* to me; the only Rocky Mountains I had seen were on paper, the only Grand Canyon, a photograph in a book.

The Great Plains did their best to entertain me, putting on earnest displays of corn, soybeans, and wheat. I counted combines to stay alert, while the Chevy smoked oil to pass the time. Then Colorado's Front Range soared up on the horizon. I had never seen anything like it. Raised in the East and schooled in the Midwest, the biggest thing I had ever laid eyes on was a Silurian reef. "Hey," I thought, "now that's a *mountain!*" Soybeans fell behind. The air thinned and cooled. Dark pines cascaded down the slopes. Higher up, the pines thinned, revealing naked rock bent into grotesque contortions—testimony to the forces that pushed up the Rockies some sixty million years ago. Over the next days, every bend in the road brought something magnificent into view. Crossing the Rocky Mountain roofline of the continent, I traversed a world that seemed closer to the sky than to the Earth, where the lowest valleys lie higher than the highest peaks back east. I followed the Colorado River from its source across the broad plateau that bears its name—a great platform, improbably high, clawed by red-rock canyons of otherworldly beauty. Dropping off the Colorado Plateau, I entered the Basin and Range

Province, a landscape rhythmically cleaved by north–south trending mountain ranges that rise in immense silence from broad, gravelly basins. Nature's rules seemed reversed here. In the East, rivers get bigger downstream; here, they shrank and disappeared downstream, sucked up by the bottomless aridity. In the East, trees eventually end as you go *up* the mountains. Here, the trees ended as I came *down*, for in the arid West the mountains poke like wooded islands out of seas of desert air.

Onward I went, delighting in the grand, implacable emptiness of it all. "Beyond the wall of the unreal city," Edward Abbey wrote, "there is another world waiting for you. It is the old true world of the deserts, the mountains, the forests." I found that old true world in the American West. Civilization, which has paved or tilled over most of the nation east of the Rockies, here lay scattered in towns and cities separated by vast, wild spaces. "In God's wildness lies the hope of the world," John Muir declared. And best of all (with apologies to Muir, who spent much of his life defending the world's biggest plants), there were few plants in this arid sector of the nation—in other words, not too much annoying biology to obscure the rocks, the database of the Earth's history.

I didn't know it then, but that first trip across the American West would reset the arc of my life. Through the years that followed—through graduate studies, tenure, grants, research papers, and the annual rhythms of teaching—the West closed its grip on me. Now, the thinnest of excuses to go to the mountains have given way to no excuses at all. With moves so familiar that I could do them in my sleep, I carry my gear from the garage to the truck, kiss my wife goodbye (unless she's decided to come along), and head once again into the wild, pulled there by the hooks that the West sunk into me long ago.

Unfold a map of North America and the first thing to grab your eye, probably, is the bold shift between the Great Plains and the Rocky Mountains. East of that line, the land is mostly smooth. To the west, it is raw and rough-hewn—a heap of twisted rock shoved skyward in the last few moments of geologic time. This book takes a trek across that rough side of the continent to tell the story of its rocks, rivers, and mountains and how they came to be. I tell it in the way that comes naturally to me as a field geologist—by taking you directly to the outcrops and landscapes that tell the story. The Earth's history lies outdoors, in the mountains and deserts, under the wind and sky; you won't find it under a roof. But your participation doesn't require getting sunburned or sweaty, because all of the outcrops are also here, between these pages. If at some point you find yourself wanting to get closer to the geology, to see the evidence firsthand, appendix II gives the locations of most of the sites and sights that we'll visit.

The book follows a west-to-east traverse across the central portion of the western United States. We begin on the California coast near San Francisco and end on the Great Plains east of Denver. I chose to go west to east because I think that direction—from the continent's edge inward—works best for the geologic story. Our traverse coincides roughly with the route of the nineteenth-century transcontinental railroad and with today's Interstate 80. But the book is not a road guide. It takes a broader view and meanders widely, angling south to Arizona's Grand Canyon and north to Idaho's Snake River Plain as needed to fill in the details. I chose a central traverse across the West (as opposed to a more southern or northern route) because I think the central region tells the most interesting and comprehensive stories of how the West came to be. With few exceptions, my focus is on developments of the last one hundred million years. That's a small slice of geologic time (if the Earth were one year old, it would represent the last eight days), but it encompasses most of the events that have shaped the West that we see today.

Science moves forward on two legs, and one can't lag too far behind the other if progress is to be made. The first leg consists of ideas about how nature works. The second comprises the data and technologies that allow us to test those ideas against nature's evidence. You might think that the first leg would lead the way, but it doesn't always. In geology, new technologies have often led the way to new discoveries. Techniques of precision surveying, developed mostly during the nineteenth century, allowed geologists to accurately map the contorted rocks of the West and reconstruct their history, somewhat like flattening a crumpled newspaper and putting the torn pieces back together. Radiometric dating, developed during the first half of the twentieth century, revealed the ages of rocks using the radioactive clocks inside. Complementing this came painstaking cataloging of fossils, which paid off with information about the ancient environments of the West and the ages of the rocks in which the fossils are found. Then came seismic receivers, which, by picking up tremors from distant earthquakes, gave us X-ray-like pictures of the once-invisible world of crust and mantle deep beneath our feet. Then came the Global Positioning System (GPS), which allows us to measure how the West moves and grows today—for as you'll see, much of the West is a living landscape, reshaping itself *right now* in response to the demands of heat escaping from the guts of the Earth.

If I can boil nearly two centuries of geologic exploration in the American West down to one essential point, it is this: the West is rough, mountainous, and geologically alive because our continent is moving. Two hundred and fifty million years ago, North America was locked in a firm embrace with the rest of the world's continents, forming a single gargantuan landmass called Pangaea. Then it tore free and headed west, opening the Atlantic Ocean in its wake. On its westward journey, the continent overrode several large sections of the ancient Pacific Ocean floor. Three big things happened in the West as a result. First, the continent grew substantially as ancient islands and bits of old seafloor collected against its leading edge like debris against a broom. Second, mountains blistered upward from Alaska to Central America to build the North American Cordillera—the great belt of mountains that stretches from Alaska to Panama along North America's western edge. Third, vast quantities of molten rock welled up throughout the West, spewing volcanic fire across the land while seeding the deep crust with precious metals. We'll revisit these themes—continental growth, mountain building, volcanism, and precious metal formation—throughout the chapters ahead.

"History is all explained by geography," Robert Penn Warren once remarked, to which he might have added, "and geography is all explained by geologic forces." In the wake of the exploratory expeditions of Lewis and Clark (1804–1806) and John C. Frémont (1842–1846), Americans in increasing numbers sought out lands and livelihoods west of the Rockies. More often than not, geology determined why they went and where they settled. Some sought wealth—gold and silver in particular—from metal deposits precipitated during the continent's westward journey. Others sought land on which to ranch or farm. In so doing, they laid out their communities and irrigation systems in ways dictated by the rivers and mountains, conforming by necessity to landscapes put in their way by the continent's westward migration. Today, we depend on those same mountains to trap and dole out the West's most precious resource—water. California could not support its population or its multi-billion-dollar agricultural economy if not for water gathered (as snow) by the Sierra Nevada and the Rocky Mountains. In fact, California owes its very existence to the continent's westward migration. Before North America started west, there was no California (or Oregon, Washington, western British Columbia, or western Mexico, for that matter). Bits and pieces of old ocean floor collected against the prow of the advancing continent to assemble the Golden State—and plant gold there, too.

There is much to tell about this jagged young land. If this book has a purpose beyond what it has given me—the carefree road trips, the fun of new discoveries, the halcyon nights under star-spangled skies—it is to inform the wonderment that everyone feels when confronted with the spectacular geologic world of the American West.

PART I

CALIFORNIA

For an extremely large percentage of the history of the world, there was no California. . . . Then, a piece at a time—according to present theory—parts began to assemble. An island arc here, a piece of a continent there—a Japan at a time, a New Zealand, a Madagascar—came crunching in upon the continent and have thus far adhered.

JOHN MCPHEE, 1998

The formula for a happy marriage? It's the same as the one for living in California: when you find a fault, don't dwell on it.

JAY TRACHMAN

We learn geology the morning after the earthquake.

RALPH WALDO EMERSON, 1883

1

GOLDEN GATE

The Farallon Islands poke like rotten teeth out of the Pacific Ocean thirty miles west of San Francisco. The islands get in the way of deep currents, forcing cold, nutrient-rich waters to the surface. Like garden fertilizer, the nutrients—iron, nitrate, and phosphate mostly—trigger blooms of tiny plant-like phytoplankton. Life converges on the plankton. Seabirds swarm the skies. Seals and sea lions crowd the shoal waters. Hunger pulls them toward the sea, but fear keeps them near the shore. The fat sea mammals form a 24/7 snack stand for Great White sharks. Boatloads of tourists motor out from San Francisco to bob on the swells, scanning the ocean with binoculars. If they're lucky, they'll have a *Discovery Channel* moment—the one where a 2000-pound shark erupts from the water with a seal twisting in its bloody jaws.

The Farallon Islands are made of granite. On the face of it that seems a dry fact, for granite is an exceedingly common rock. It forms much of the basement rock of the continents and the uplifted cores of many continental mountain ranges. But *continental* is the key word here. Granite is a continental rock. Finding it on an oceanic island is strange. You could spend your life wandering around Hawaii or Tahiti or most of the other islands of the Pacific and never find a speck of granite. What is a continental rock like granite doing out at sea on the Farallon Islands?

The unexpected answer comes when we compare the islands' granite to granite from mainland California. The granite of the Farallon Islands, it turns out, is nearly identical in age and chemistry to granite found in the southern Sierra Nevada—300 miles southeast and nearly 100 miles inland from the California coast. The Farallon

Islands, in other words, appear to be pieces of the Sierra Nevada that have split away from the continent and wandered out to sea. The reason is the San Andreas fault. Starting about eighteen million years ago, motion along the fault snipped off a large piece of the southern Sierra Nevada and slid it northwest toward San Francisco. Salinian Block is the name given to this chunk of dispossessed Sierra Nevada granite. The block includes the Farallon Islands and much of coastal California north and south of San Francisco. If present trends continue, the Farallon Islands, riding with the Salinian Block, will arrive at the coast of Alaska in about seventy million years.

For many people, the notion of a big chunk of the southern Sierra Nevada presently offshore of San Francisco and drifting toward Alaska falls somewhere between disturbing and impossible. It's a fair reaction. It comes from viewing the Earth through the lens of human time—a perspective that gives the illusion of stability to a world that, over geologic time, is radically mobile. A flash of a strobe light on a dance floor gives the same illusion. The dancers in that microsecond appear frozen in place, even though they're actually in continual motion. In human time, the continents and ocean basins appear fixed and permanent. To know that the Atlantic Ocean is forty-three feet wider today than when Columbus crossed it lifts, just a little, the veil of our illusion. We need to tear that veil away to see the Earth as it really is. When we do that, we see a different planet. Continents don't stay put. They drift like loose barges across the face of the Earth. Where they tear apart, new ocean basins open. Where they crunch together, ocean basins disappear and mountains rise in their place.

This action happens because the Earth's outer rind is split into moving plates of rock (frontispiece figure). The plates range fifty to one hundred miles thick, and several hundred to several thousand miles across. They move, in various directions, a few inches each year (figure 1.1). Continents are not plates, but they form parts of some plates. The continents ride like passengers on the plates, going where the plates take them, rearranging the face of the Earth over geologic time. Just about every major geologic feature on Earth—including volcanoes, earthquakes, mountain belts, metal and mineral deposits, and the large-scale features of the ocean floor—results from the movements of these plates. This discovery, called plate tectonics, stands at the pinnacle of human insight, shoulder-to-shoulder with biological evolution, the structure of the atom, quantum physics, and the expansion of the universe. Its power undergirds nearly everything that we'll explore in this book. The Earth's moving plates have written the story of the American West.

· · ·

As you leave the Farallon Islands and ride the swells back to San Francisco, the first piece of the mainland to loom out of the fog is usually Point Bonita, a small peninsula that pokes like a crooked finger into the Golden Gate. A lighthouse at the tip of the point warns ship captains to steer clear of the dark cliffs and skerries that shatter

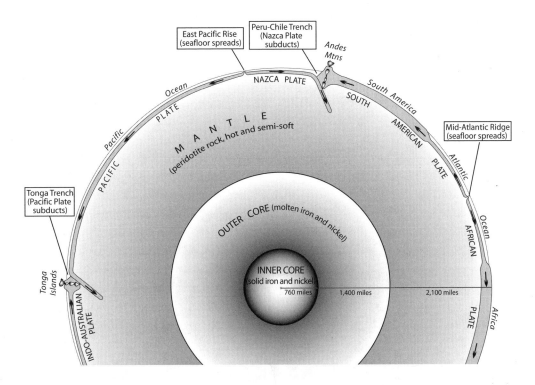

FIGURE 1.1

A schematic slice through the Earth bisecting the Indo-Australian, Pacific, Nazca, South American, and African plates. Plates grow apart from one another at mid-ocean ridges through seafloor spreading and come together at ocean trenches, where one plate dives beneath the other in the process of subduction. For clarity, the diagram greatly exaggerates the thickness of the plates, the depth of the oceans, and the topographic relief of the land surface. (If I had drawn the diagram to scale, the ocean would be thinner than the thinnest line, and the Andes and the Tonga Islands would be near-invisible bumps.)

the Pacific swells to white. The rock of Point Bonita is not granite but pillow basalt: greenish-black volcanic lava taking the form of pillow-sized blobs stacked upon one another like sacks of grain (figure 1.2).

The pillow basalt at Point Bonita and the granite of the Farallon Islands are utterly different rocks; you would more easily confuse Martin Luther King, Jr. and Eleanor Roosevelt. But they have one thing in common—they have both traveled a long way. In fact, the Point Bonita pillow basalt has traveled *several thousand miles* to end up here at the Golden Gate—a tectonic journey that makes the wandering granite of the Farallon Islands seem a homebody by comparison. The pillow basalt testifies to an epic tale of crustal mobility, powered by the Earth's moving plates. To understand this story, we need to look at how the Earth creates and destroys its ocean floor.

Pillow basalt (also called pillow lava) forms from lava eruptions on the deep seabed at mid-ocean ridges—broad undersea mountain belts that wrap around the planet like

FIGURE 1.2

View from the air looking east toward Point Bonita in the Marin Headlands, with the Golden Gate
Bridge beyond (top). All around the point are superb exposures of pillow basalt—heaped blobs of dark
volcanic basalt (bottom) that once formed part of the deep ocean floor several thousand miles west of
California.

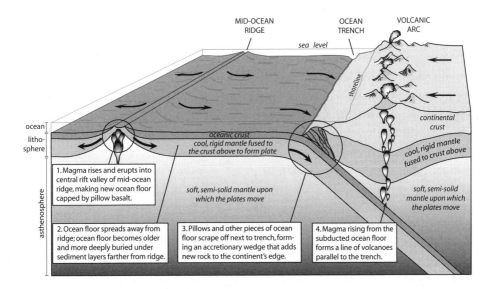

MID-OCEAN RIDGE OCEAN TRENCH VOLCANIC ARC

sea level

shoreline

ocean litho-sphere

asthenosphere

oceanic crust
cool, rigid mantle fused to the crust above to form plate

continental crust

cool, rigid mantle fused to crust above

1. Magma rises and erupts into central rift valley of mid-ocean ridge, making new ocean floor capped by pillow basalt.

soft, semi-solid mantle upon which the plates move

soft, semi-solid mantle upon which the plates move

2. Ocean floor spreads away from ridge; ocean floor becomes older and more deeply buried under sediment layers farther from ridge.

3. Pillows and other pieces of ocean floor scrape off next to trench, forming an accretionary wedge that adds new rock to the continent's edge.

4. Magma rising from the subducted ocean floor forms a line of volcanoes parallel to the trench.

FIGURE 1.3

The movements of the Earth's plates center on two opposing processes: seafloor spreading and subduction. Seafloor spreading at mid-ocean ridges forms new ocean floor, capped by pillow basalt. Subduction at ocean trenches destroys the ocean floor, except for pieces that scrape off next to the trench to form an accretionary wedge. For clarity, the diagram greatly exaggerates the vertical topographic relief of the ocean floor and the land surface.

the seams on a baseball. Deep rift valleys run down the centers of these ridges. The ocean floor slowly splits at these rifts, and lava wells up into the gap to congeal like blood in a wound. The lava squeezes out as toothpaste-like gobs, glowing red for an instant before the cold seawater quenches it into mounds about the size of a bed pillow. As the seawater circulates through cracks in the hot rock, chemical changes trigger the growth of chlorite minerals, which tint the black basalt green.

Take a submarine down to the central rift valleys of the Mid-Atlantic Ridge or the East Pacific Rise, and you can see the seabed splitting and the lava erupting and forming pillows. Cruise along the deep seabed in either direction away from these central rifts, and the pillows become more deeply buried in seabed muck. That muck—technically known as ooze—is made mostly of microscopic plankton remains. Geologists, by puncturing the deep seabed with thousands of drill holes, have learned two important things about this ooze and the pillow basalt underneath it. First, drill down through enough ooze and you eventually hit pillow basalt. Second, the farther you go from a mid-ocean ridge, the thicker the ooze and the older the pillow basalt beneath it.

These facts tuck neatly into the concept of seafloor spreading—the Earth's system for manufacturing new ocean floor (figure 1.3). After lava erupts at a mid-ocean ridge to form pillows, it makes room for new lava by spreading away in opposite directions, like two oppositely moving conveyor belts. As the seabed spreads from the ridge, planktonic ooze slowly buries the pillows under ever-thicker layers. The world's mid-ocean ridges spread in this conveyor belt-like manner about two inches per year, on average.[1] That's a plodding pace in human time, to be sure—but so what? Compared to geologic time, human time is virtually insignificant, like a couple of still-frames in a feature-length movie. Take two inches per year over the entire 40,000-mile-long mid-ocean ridge system and watch what happens when geologic time takes over. In 2.4 million years, the Earth will manufacture an area of ocean floor equal to that of the lower forty-eight U.S. states. And in 158 million years—a span less than 4 percent of the age of the Earth—that pace of seafloor spreading will create an area of ocean floor equal to the surface area of the entire planet![2]

What happens to all of this newly minted ocean floor? With the seabed constantly splitting and growing from its mid-ocean ridges, you might expect the planet to be swelling like a balloon. But it's not, because the ocean floor is destroyed apace with creation. This happens at ocean trenches—great troughs where the ocean floor bends down beneath a neighboring plate to plunge into the Earth's hot interior (figure 1.3). This process—called subduction—consumes the ocean floor almost as fast as seafloor spreading produces it.

Almost.

Subduction, it turns out, doesn't eat up every square foot of ocean floor churned out by seafloor spreading. There is a small—but crucial—imbalance between the creation of the ocean floor and its destruction. That imbalance has created California. It explains (among other things) how pillow basalt has ended up at Point Bonita in the Golden Gate.

As the seafloor slides into an ocean trench, the edge of the adjacent plate can act like a plow blade, scraping off slivers of seafloor rock. Anything sticking up from the seafloor—submarine plateaus, seamounts, or even entire islands—can get scraped off as well. If a continent tries to follow the ocean floor into a trench, it won't go far.[3] Instead, after going a little way down the trench, the edge of the continent may lever a piece of ocean floor up out of the sea like a pig wedging a truffle out of the ground. The astonishing result—played out over the scope of geologic time—is that *continents grow bigger as pieces of oceanic rock collect against their edges*. Stand anywhere in western Mexico, California, Oregon, Washington, British Columbia, or Alaska, and you stand—more often than not—on pieces of former ocean floor that have been scraped or lifted off the deep seabed and marooned on North America's western edge, mostly during the last 200 million years (figure 1.4).

The reason this has happened comes down to North America's long history of westward migration, combined with the eastward migration of pieces of old ocean floor that have slid underneath or collided with North America's western edge. Two hundred and fifty million years ago, our continent was firmly stuck to Eurasia and Africa as part of the supercontinent Pangaea. But about 200 million years ago, North America began to tear away and head west. New ocean floor created by seafloor spreading in the young, growing Atlantic had to be balanced by destruction of ocean floor elsewhere. That hap-

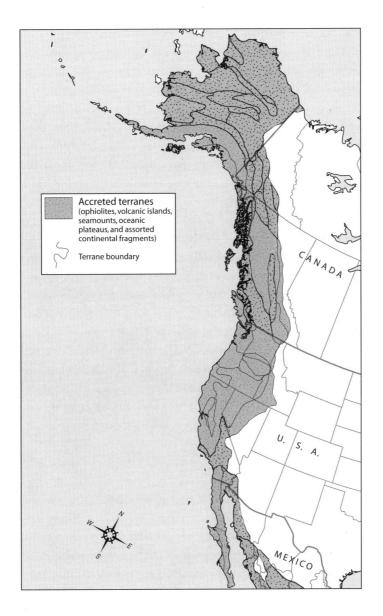

FIGURE 1.4

The growth of western North America. The western edge of the continent comprises vast strips of immigrant rock, called accreted terranes, brought here by subduction from far out in the ancient Pacific Ocean. Before they docked onto the continent, these terranes were pieces of oceanic crust, volcanic islands, seamounts, oceanic plateaus, and fragments of small continents. Like groceries piling up at the end of a checkout-line conveyor belt, they have collected, one behind the other, against the continent's western edge, mostly during the last two hundred million years. We distinguish individual accreted terranes by the great faults that separate them and the far-flung rocks and fossils within them.

pened as the continent overrode or collided with multiple pieces of ancient Pacific Ocean floor. Much of this old seabed went under the continent, but several million square miles of it glommed onto the continent's western edge, adding the new geologic real estate shown in figure 1.4.

Now we can circle back to Point Bonita and shine the light of these concepts on the pillow basalt there. More than one hundred million years ago, the pillows formed part of the seabed of the Farallon Plate—a big oceanic plate that once plunged under North America's western edge. Instead of sliding down the trench with the rest of the plate, the pillow basalts that we now see at Point Bonita sheared off next to the trench to join what is called an accretionary wedge. Jammed against the continent, the rocks of the accretionary wedge became new land for California (figures 1.5 and 1.6).

The pillow basalt at Point Bonita is just the tip of this accretionary rock iceberg. Pillow basalts crop out throughout the Marin Headlands above Point Bonita, on the Twin Peaks near downtown San Francisco, high above sea level on Mount Diablo east of Berkeley, and in dozens of other places throughout the Bay Area. What city but San Francisco could feature a tourist guidebook called *A Streetcar to Subduction*, which will take you to outcrops of these deep-sea rocks by way of public transportation? Franciscan Complex is the name given to this collection of deep-sea rocks that collected in the accretionary wedge above the Farallon Plate. You'd be hard-pressed to find a more mixed-up, fault-shattered mess of rock on Earth. The Franciscan Complex is what happens when subduction takes quadrillions of tons of old seabed and jams it against the edge of a continent.

·　　　·　　　·

Above Point Bonita and its pillow basalts, the land rises steeply in mottled shades of coastal scrub to form the north shoulder of the Golden Gate. These are the Marin Headlands, a piece of near-wild California, where raptors, bobcats, mountain lions, and coyotes far outnumber people. Grim concrete relics of military readiness dot the headlands: gun batteries from two world wars and a surface-to-air missile battery from the Cold War. When active, the gun batteries could fire 2100-pound shells as far as the Farallon Islands. Most of the old gun batteries on the Marin Headlands sink their foundations not into pillow basalt, but chert—another abyssal import scraped off the Farallon Plate.

After pillow basalts form and spread away from mid-ocean ridges, they catch a steady rain of planktonic debris (ooze) from the waters above. Many plankton cells contain lacy microscopic skeletons made of calcium carbonate or silica. The calcium carbonate skeletons often dissolve under the pressure and cold of the abyss, but the silica ones usually don't. Instead, they settle like fine snow onto the pillowed seabed, where they eventually harden into chert—a siliceous rock that breaks in scalloped curves like thick glass. Chipped by skilled hands, chert becomes arrowheads and

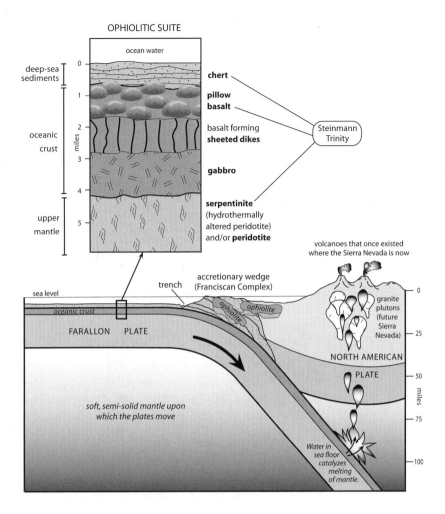

OPHIOLITIC SUITE

FIGURE 1.5

An east–west cross section portraying the Farallon and North American plates about one hundred million years ago. Magma, coughed up from the subducting Farallon Plate, rises and erupts as arc of volcanoes where the Sierra Nevada is today. Magma that doesn't make it to the surface solidifies underground to make granite that would later rise to form the Sierra Nevada range (see chapter 4). To the west, in what is now the San Francisco Bay area, pieces of the Farallon Plate scrape off and collect into an accretionary wedge: a fault-slivered mass of displaced deep-sea rock. The remains of this hodgepodge today comprise a geologic unit known as the Franciscan Complex. The three rocks of the Steinmann Trinity—chert, pillow basalt, and serpentinite—occur abundantly within the Franciscan Complex, particularly around the Golden Gate.

A complete *ophiolitic suite*—a vertical slice through the oceanic crust and upper mantle, as shown in the top part of the figure—comprises a five-fold grouping from top to bottom of (1) chert and/or other deep-sea sediments, (2) pillow basalt, (3) sheeted dikes of basalt (representing magma injected into fractures in the spreading ocean floor below the erupting pillows), (4) gabbro (rock from the same magma that made the basalt pillows and sheeted dikes but which cooled slowly deep in the ocean crust), and finally (5) serpentinite and/or peridotite (the rock of the Earth's mantle below the oceanic crust).

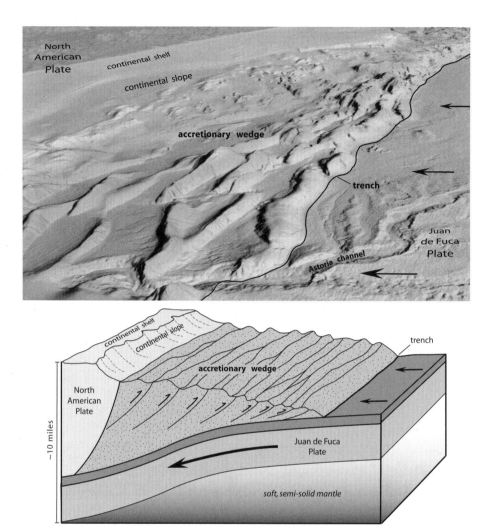

FIGURE 1.6

This accretionary wedge, forming today off the coast of Oregon, represents a modern analog to how the Franciscan Complex in the San Francisco Bay area formed. The wedge is accumulating from layers of sediment scraped off the Juan de Fuca Plate, a tail-end remnant of the Farallon Plate that subducts today beneath southern British Columbia, Washington, Oregon, and northern California. The image of the seafloor reconstructed from sonar images (top) looks obliquely south across the accretionary wedge with the ocean removed. Notice the "rumpled-towel" appearance of the seafloor. The block diagram (bottom) illustrates the inferred pattern of faults and folds formed as layers of seabed rock scrape off the subducting plate to join the growing accretionary wedge.

FIGURE 1.7

About 160 million years ago, these chert beds lay several miles below sea level in the tropical Pacific perhaps two thousand miles from California. Originally flat lying, they bunched up like wet laundry as they scraped off to join the accretionary wedge forming above the subducting Farallon Plate. The exposure lies along Conzelman Road in the Marin Headlands. Note bored spouse for scale.

spear points. (Flint is chert in archeological clothing.) When subduction scrapes pillows off the seabed, the chert above the pillows comes along for the ride. The chert layers often scrunch into fantastic contortions as they mash into the accretionary wedge (figure 1.7).

The chert in the Marin Headlands comes mostly from microscope creatures called radiolarians, a common form of siliceous plankton. The radiolarians date from the Early Jurassic to Middle Cretaceous time periods, meaning from about 200 million to 100 million years ago. (See the geological time scale at the front of the book.) Most of the radiolarians found in the Marin Headlands are of tropical species that lived in the warm equatorial waters of the ancient Pacific. Think about what this means—the chert that we now see in the Marin Headlands must have traveled all the way from the equator, *some 2,000 to 3,000 miles,* before subduction added it to California's doorstep. Most of California is built of equally well-traveled rocks.

My favorite stop in the Marin Headlands is Battery 129, an abandoned gun battery perched on bluffs high above the Golden Gate. Battery 129 draws tourists like bears to honey; the view of the Golden Gate Bridge, and of downtown San Francisco beyond, is the best that can be had outside of a helicopter ride. But forget the view, I tell anyone who will listen. A much more dramatic sight lies in the nearby road cut. If you go to the Marin Headlands, don't miss the chance to put your nose on the road cut by Battery 129 (figure 1.8). The cut exposes a cross section of the Jurassic ocean floor, where layers of brick-red chert can be seen lying directly on top of pillow basalt. Lay your hands on these rocks and time shifts to 160 million years ago, to the floor of the tropical Pacific, three miles below sea level and more than 2,000 miles from California. The weathered and crumbling pillows under your hands are now fresh lava, squeezing like incandescent toothpaste out of the splitting seabed, glowing for an instant before the cold water quenches them to stone. Move upward now and lay your hands on the chert. Now billions of microscopic radiolarians are settling through the dark waters to bury the seabed in a rising tide of siliceous ooze. All the while, the conveyor belt-like movement of the Farallon Plate is carrying the seabed northeast, toward the trench that once lay off California's coast. There, the seabed bends like a down escalator into the trench. The chert and pillow basalt seemed destined to disappear into the bowels of the planet—but they don't. Instead, they shear off to join the great accretionary wedge above the Farallon Plate. Crowded and squeezed upward by other rock wedging in from behind, they rise slowly out of the sea. And so, here they are today, remarkably unmangled considering their incredible journey. You can see and touch all of this history in this one outcrop—one that captures a span of time nearly one million times longer than the history of the United States.

·   ·   ·

Turning away from Battery 129, I stand with other tourists looking down into the Golden Gate. Four times a day, the Moon tugs an average of 436 billion gallons of seawater (enough to fill about 1,100 football stadiums) back and forth through the three-and-a-half mile-wide strait, where the tidal waters churn with powerful currents and eddies. Engineers in the 1930s, building what was then the world's longest suspension bridge, had to contend with these currents. And they faced a larger problem: how to anchor the two colossal concrete piers that would hold up the twin towers of the Golden Gate Bridge.

For the north pier, the engineers dug into the bedrock below the Golden Gate and found hard, competent chert and pillow basalt. No worries there—support for the north pier was assured. The south pier was more problematic. The bedrock on the south side of the Golden Gate is mostly a soft, slippery rock called serpentinite. The south pier would have to stand on this soft rock, in powerful tidal currents, within two miles of the San Andreas fault, which, in 1906, had unleashed the great earthquake that leveled most of San Francisco.

To make a foundation for the south pier, the engineers decided to hollow a cavity out of the serpentinite, 110 feet deep and an acre across, and fill it with concrete like a dentist packing a root canal. Such a footing would still need to sit on reasonably competent rock, however; otherwise, it would be like pouring concrete into a mold of butter. In December 1934, the engineers finished an inspection well that pierced deep into the

original surface of the ocean floor before the chert layers accumulated

tilted beds of RED CHERT (radiolarian remains that accumulated on the pillow basalt)

cyclist unaware of profound geology

PILLOW BASALT (lava formations of the deep ocean floor, now deeply weathered)

FIGURE 1.8

This road cut near Battery 129 in the Marin Headlands exposes a slice of mid-Jurassic (roughly 160-million-year-old) ocean floor, from the pillow basalt of the oceanic crust up through the layers of brick-red chert that accumulated on top of the pillows. Fault movements have tilted the whole sequence toward the west (left).

serpentinite below the strait. The highlight of the effort came as the distinguished Berkeley geologist Andrew Lawson was lowered into the well. San Francisco held its collective breath as Lawson whacked away with his rock hammer deep in the well. Lawson emerged and pronounced the serpentinite sound. "The rock of the entire area is compact, strong serpentine remarkably free from seams of any kind," he reported. "When struck with a hammer," he added, "it rings like steel." San Francisco cheered. The city would have a Golden Gate Bridge.

It's hard to reconcile what Lawson found with what you can see today in the blue-green bluffs near Fort Point, where the Golden Gate Bridge takes off from San Francisco (figure 1.9). Here, as little as 300 yards from the bridge's south pier, the serpentinite is about the worst excuse for a rock that I've ever seen—riven with fractures, slick as soap, and soft enough to carve with your fingernail. Anyone who visits these bluffs and feels the serpentinite disintegrating between his on her fingers could be forgiven for taking a sudden interest in the Golden Gate ferry schedule. Granted, the bridge has stood up well to earthquakes since it opened in May 1937. It has yet to face the ultimate test, however—a repeat of the 1906 San Andreas earthquake.

FIGURE 1.9

An outcrop that frequent users of the Golden Gate Bridge should avoid if they want a good night's sleep. The view (top) looks north from Baker Beach toward Fort Point along bluffs of blue-green serpentinite, the same rock that holds up the bridge's south pier. The serpentinite in these bluffs is an unholy mess: spongy, slippery, and riddled with fractures, many of which are filled with veins of asbestos—the whitish laceworks in the close-up (bottom). Serpentinite is peridotite (rock from the Earth's upper mantle) altered by hydrothermal reactions with seawater near spreading mid-ocean ridges.

Like the chert and pillow basalt we've already seen, the serpentinite around the Golden Gate was once part of the Farallon Plate, and it, too, was expelled from the oceanic abyss in the process of assembling California. Serpentinite begins as peridotite—the rock of the upper mantle just below the oceanic crust (top diagram in figure 1.5). Peridotite is normally a robust, crystalline rock that sparkles with sea-green olivine and dark-green pyroxene minerals. But where the seabed splits and spreads at mid-ocean ridges, seawater can percolate down to the uppermost mantle and transform the peridotite there into serpentinite. Water molecules, backed by extreme heat and pressure, invade the atomic lattices of peridotite minerals, converting them into soft, hydrated forms like antigorite, chrysotile, and—my nomenclatural favorite—lizardite. These are the so-called serpentine minerals, named for their snakeskin-like shades of mottled green. Serpentinite is rock made up of serpentine minerals.

The German geologist Gustav Steinmann, working in the Alps in the late nineteenth century, was the first to recognize that chert, pillow basalt, and serpentinite form a trinity. Where you find one, he discovered, you often find the other two nearby. In 1892, Steinmann visited San Francisco, where a young Andrew Lawson (forty-two years before he would be hammering on serpentinite at the bottom of the south pier well) took Steinmann to the then-bridgeless Golden Gate. Steinmann had déjà vu. The rocks here, he told Lawson, matched exactly those that Steinmann had studied in the Alps.

Steinmann's Alpine work had shown him that (where faults had not confused matters too badly) the trinity of chert, pillow basalt, and serpentinite always occurred in a predictable order: chert on top, pillow basalt in the middle, serpentinite on the bottom. Since chert forms mostly on the deep seabed, Steinmann reasoned that the pillow basalt and serpentinite could only have come from even *farther* down. In other words, he realized, the trinity must represent pieces of the deep ocean floor, from the capping sediments down through the oceanic crust and into the mantle beneath. Steinmann dubbed these consistent trinities "ophiolites," based on their abundant serpentine minerals (*ophis* = "snake" in Greek). In 1905, Steinmann published a definitive paper on ophiolites. Ever since, the three-fold association of chert, pillow basalt, and serpentinite has been enshrined in geology's lexicon as the Steinmann Trinity (figure 1.5).[4]

Although Steinmann correctly deduced that ophiolites represent pieces of the deep ocean floor that are now on land, he went to his grave puzzled as to *how* these deep-sea rocks made the journey from the abyss up onto land. Ophiolites remained a conundrum until 1969, when California geologist Eldridge Moores—his mind a-fizz with the fresh ideas of plate tectonic theory—demonstrated how ophiolites could be dislodged from the ocean floor during subduction and plate collisions. Suddenly, ophiolites fit neatly into all sorts of interesting problems. Pillow basalt in the mountains of Tibet? Trapped like baloney in a sandwich when India crunched into southern Asia to shut the Tethys Seaway and push up the Himalayas. Serpentinite in the Alps and Apennines? Pushed out of the sea when Africa closed with Europe, collapsing the Tethys Seaway to form today's Mediterranean. A ten-mile-thick slab of ocean floor in the Oman desert?

Thrust onto the Arabian Peninsula as it closed with Asia. The Steinmann Trinity at the Golden Gate? Slabs of deep seabed scraped off the subducting Farallon Plate.

In 1965, the California legislature chose serpentinite as the official state rock. The winds of the plate tectonic revolution had just begun to stir in the halls of academe. No one yet knew that serpentinite, so emblematic of California, isn't native to the state. Arguably, *no* rock is native to California since virtually the entire state is built of imported geology. (I'm ignoring here the veneer of locally derived sediments and lava layers that sometimes cover the bedrock.) Wherever you stand in this state, if your feet are on bedrock, the odds are that you're standing on an immigrant, reeled in by subduction from the far reaches of the Pacific in the process of assembling California.

NOTES

1. The slowest mid-ocean ridges spread about a half-inch per year, whereas the fastest ridge—the East Pacific Rise—spreads more than six inches per year in some areas.

2. Here's the math to make the point. Take the average ocean floor spreading rate of two inches per year, multiply that by the 40,000-mile-long mid-ocean ridge system, and you get a global production rate of 1.25 square miles of new ocean floor per year. Let's assume that this is a typical rate over geologic time (although spreading rates were likely a bit faster when the Earth was younger and hotter). We can use this value to calculate how long it would take the Earth to make certain amounts of new ocean floor. For instance, the area of the lower forty-eight U.S. states is about three million square miles. Seafloor spreading would make this amount of ocean floor in 3×10^6 square miles $\div 1.25$ square miles per year $= 2.4$ million years. The surface area of the Earth is about 197 million square miles. Seafloor spreading would make this amount of ocean floor in 197×10^6 square miles $\div 1.25$ square miles per year $= 158$ million years. Since the Earth is about 4.56 billion years old (see appendix I for that story), it's fair to say that a truly stupendous amount of ocean floor, several times greater than the surface area of the Earth itself, has been created and destroyed over geologic time.

3. Continents can't be subducted far because the low-density rock of the continents makes them buoyant in the mantle. Visualize trying to push a big slab of Styrofoam down to the bottom of a swimming pool; you can get it part way down (think of India jamming under Tibet to hoist the Himalayas), but its buoyancy soon resists further subduction. In contrast, the high-density rock of the ocean floor readily sinks into the mantle.

4. Ironically, Gustav Steinmann—the father of the ophiolite concept—may have never seen a complete ophiolite. In the Penninic zone of the Alps where he worked, the ocean floor sequence is incomplete, consisting only of his famous trinity. Studies of better-preserved ophiolites, particularly in Oman, Cyprus, and California's Coast Ranges, have shown that a complete section through the ocean floor—known as the "ophiolitic suite"—generally comprises a five-fold grouping, as shown in figure 1.5.

2

MOTHER LODE

As when some carcass, hidden in sequestered nook, draws from every near and
distant point myriads of discordant vultures, so drew these little flakes of gold
the voracious sons of men.

HUBERT HOWE BANCROFT, 1888

Beginning in 1848, as the news of gold in the western Sierra Nevada foothills burst upon the world, people from all corners of the Earth began to converge on California. It was, in the words of historian H. W. Brands, "the most astonishing mass movement of peoples since the Crusades."

Thousands came overland from the eastern United States, following the hard wagon roads across the American West. Thousands more came by ship. As the great brigs, barks, and schooners dropped anchor in San Francisco Bay, the crews abandoned ship as fast as the passengers, all heading off to the gold fields. "Keep clear of there [San Francisco] by all means," an alarmed Nantucket whale ship owner warned his captain, "or you will not have a crew to bring you out." By the early 1850s, hundreds of vessels lay rotting in the bay—a nautical black hole where ships went in but not out. Lumber from the deserted ships fed the explosive growth of San Francisco. Masts became pier pilings; planks became new shops and houses.

Communities nationwide watched with alarm as young men abandoned the lives expected of them and headed to California. "The gold mania rages with intense vigor, and is carrying off its victims hourly and daily," the *New York Tribune* wailed in December 1848. "The whole country from San Francisco to Los Angeles, and from the sea shore to the base of the Sierra Nevadas, resounds with the sordid cry of 'gold! gold, GOLD!' the *San Francisco Californian* reported, "while the field is left half planted, the house half built, and everything neglected but the manufacture of shovels and pickaxes." Newspaper editors published impassioned pleas to men to stay home—and fired rebukes at their departing backs. The *Sydney Herald,* for instance, offered this advice to aspiring gold seekers:

> *What class ought to go to the Diggins?* Persons who have nothing to lose but their lives.
> *Things you should not take with you to the Diggins.* A love of comforts, a taste for civilization, a respect for other people's throats, and a value for your own.
> *Things you will find useful at the Diggins.* A revolving pistol, some knowledge of treating gunshot wounds, a toleration of strange bedfellows.
> *What is the best thing to do when you get to the Diggins?* Go back home.

The curmudgeons were largely ignored. After all, there were fortunes to be made in California for those bold enough to go. "I am coming back home with a pocketful of rocks!" forty-niner William Swain announced to his wife as he set out from New York. Edward Hargraves, arriving by schooner into San Francisco, was assured that there was "plenty of gold—for all those who will work for it."

And there *was* plenty of gold—for those lucky enough to find it. Fantastic stories floated down from the mountains: a 141-pound nugget from the Stanislaus River; a lone miner near Nevada City who washed $12,000 in gold in a single day; four men who took $9,000 each from Bidwell's Bar on the South Fork of the Feather River; a company of 30 who took $75,000 from the North Fork of the American River. Just as the occasional big payoff at a Las Vegas casino keeps hope alive (and cash flowing) among those who witness it, so the occasional big strike in California inspired thousands of men to keep digging and blasting, even as they sank further into impoverishment. Forty-niner Alonzo Delano, who traveled widely throughout the Sierra Nevada during the gold rush years, found few millionaires. But he met plenty of "disappointed and disheartened men."

> The trails and mountains were alive with those whose hopes had been blasted, whose fortunes had been wrecked, and who now, with empty pockets and weary limbs, were searching for new diggings, or for employment—hoping to get enough to live on, if nothing more. Some succeeded, but hundreds, after months and years of toil, still found themselves pining for their homes, in misery and want, and with a dimmed eye and broken hopes.

Even as the discouraging news leaked out to the world, the magnetic lure of gold continued to pull men in. "Notwithstanding the repeated failure of thousands, after months, and even years . . . it is a matter of surprise that so many should continue to leave comfortable homes, and a good business, to try their fortunes in the uncertain occupation of mining," Alonzo Delano mused. Those who avoided the mines and focused instead on supplying the miners often enjoyed surer success. A widow in Downieville sold baked pies to hungry miners. She could hardly keep up with demand, and in 1852 reported gross sales of $18,000 (about $468,000 today[1]—from selling pies!). The same year, a young entrepreneur named Leland Stanford, Jr., opened the first of several general stores. Stanford's ambition was to be the go-to man for everything a miner needed outside of female affection. "Provisions, Groceries, Wines, Liquors, Cigars, Oils & Camphene, Flour, Grain and Produce, Mining Implements, Miners' Supplies, Etc," bold letters announced above his store. Stanford proved a genius at parting miners from their money. (One winter when he ran low on vinegar—a key ingredient for flavoring miners' beans—Stanford substituted his surplus of whiskey, a move that created a tasty new dish and won him more business than ever.) Stanford's fortune would carry him to the governorship of California, endow Stanford University, and help fund the transcontinental railroad.

Hard experience taught miners a singular truth: gold is not evenly distributed in the Earth. "Even in the most auriferous sections there is only a comparatively small portion which pays the laborer abundantly," Alonzo Delano observed, "and while now and then one miner may make a good strike, by far the greater number will make scarcely day wages." Gold's fickle and uncertain distribution comes down to geology.

The average concentration of gold in the Earth's crust is about five parts per billion. Given that any rock you pick up contains several million quadrillion atoms, odds are there is some gold in every rock you find. But before you bulldoze your backyard, consider this: at five parts per billion, you'd have to process more than *200 tons* of rock to get enough gold for one wedding ring. The point is, gold mining would not exist if geologic forces didn't somehow gather up that rare gold into minable concentrations. In California, that happened in two main ways: formation of *lodes* (gold-bearing quartz veins) and formation of *placers* (gold deposits that collect in the bottoms of stream channels). Placer (rhymes with passer) gold was the first and most easily mined gold of the gold rush. But lodes proved to be a longer-lasting source. This chapter tells the story of lode gold. Chapter 3 tells of placer gold.

The formation of California's lode gold is inseparable from the story of California itself. As we saw in chapter 1, most of California has been assembled from pieces of old ocean floor stuck to North America's western edge during the last 200 million years. That process, it turns out, has everything to do with how gold came to the Golden State.

.   .   .

Gravel ricochets from braking tires as I come to a quick stop by the roadside. I'm in the western Sierra Nevada foothills, close to where Highway 20 crosses the Yuba River near the small town of Smartville. It's a 98-degree July afternoon, and not a few stares from passing motorists linger at the sight of a geologist hammering at sun-blasted rock too hot to touch. But this is not just any rock. The elephantine lumps of greenish-black that bulge from the road cut are pillow basalts—lava formations that pave the deep ocean floor, formed by seafloor spreading at mid-ocean ridges. When subduction tried to take these pillows down a Jurassic trench, they didn't go. Instead, they were wedged off the

FIGURE 2.1

Jurassic pillow basalt of the Smartville Ophiolite, in the western Sierra Nevada foothills a few miles west of Smartville. These rocks are identical to the blobby, pillow-shaped lava formations that erupt today on the deep seabed at mid-ocean ridges. Subduction brought them here—along with gold—from the far-flung reaches of the ancient Pacific.

ocean floor and left behind in the process of assembling California. The result is an ophiolite: a piece of deep ocean floor now on land (figure 2.1).

Tectonic forces have reoriented this particular ophiolite—known as the Smartville Ophiolite—in such a way that traveling east (uphill into the Sierra Nevada) mimics traveling down through the old ocean floor. (Here you might want to look back at figure 1.5 to remember what a vertical slice through the ocean floor looks like. Take that vertical slice, turn it on its side, and fold it over onto itself a couple of times, and you have the Smartville Ophiolite.) A few miles east of Smartville, the pillow basalts give way to basalt dikes. These sheet-like basalt masses, each a few inches thick, originally formed below the pillow basalts from magma that squirted upward along fractures below the mid-ocean ridge, on its way to feed the pillows on the seabed above. A few miles farther east—and thus lower in the oceanic crust—the basalt dikes give way to dark gabbro, with thumb-sized black crystals that sparkle like dark mirrors wherever my hammer breaks open a fresh surface. The gabbro congealed from the same magma that fed the basalt dikes and basalt pillows higher in the sequence. Here, though, the magma stayed deep below the mid-ocean ridge and thus cooled slowly, giving the atoms time to assemble into big crystals. My traverse through the Smartville Ophiolite eventually lands me in the town of Grass Valley. Black gabbro, the most abundant local construction stone, glistens from stone fences and retaining walls. Nearly ten million ounces of gold have come out lode mines around Grass Valley. Gold and ophiolites—there is a connection here.

Twenty miles east of Grass Valley, higher in the Sierra Nevada, I intersect another ophiolite. Mapped as the Feather River Belt, it forms a narrow band of oceanic rock that snakes eighty miles north to south through forests of oak and pine. Here my hammer, after breaking though the rock's weathered rind, exposes sea-green serpentinite and black-green peridotite—rocks from the very bottom of the ophiolitic sequence, from the mantle below the oceanic crust. It is mind-bendingly strange to see the roots of pines wedged into serpentinite—rock that once lay five miles below pillow basalts that once lay three miles below whitecaps that once billowed hundreds of miles west of where California didn't yet exist. Geology is stranger than fiction—and more interesting. According to current theory, the rocks of the Feather River Belt were pinched off the deep ocean floor in a closing vise between the then-western edge of North America and a set of volcanic islands crunching into the continent from the west.

To understand how California's lode gold came to reside in the ophiolites and other oceanic rocks of the western Sierra Nevada foothills, we need fit these rocks into the story of California's birth. The ophiolites are crucial to this story, because each one records the closing of an ocean basin and the subsequent attachment of new rock to the edge of the continent (figure 2.2). Bodies of imported rock that plate movements attach to the edges of continents are known as accreted terranes. Viewed from above, the accreted terranes of the western Sierra Nevada form three great north–south-trending belts—the eastern, central, and western belts shown in figure 2.3. Each belt comprises several terranes; the exact number is debated. The eastern belt rocks landed first, mostly before 200 million years ago. The central belt rocks crunched in behind them roughly 175 to 160 million years ago, followed by the western belt rocks from about 160 to 150 million years ago.[2] The terranes, in other words, crunched into North America in sequence, like cars in a freeway pileup—or beer cans in a beer can pileup (figure 2.4).

West of the Sierra Nevada, bedrock exposures disappear beneath the sand and gravel that fills the Great Central Valley, but seismic imaging reveals yet more accreted terranes down below. These include the Great Valley Ophiolite, which underlies much of the valley, and the Coast Range Ophiolite, whose knobby black pillow basalts and slippery serpentinites crop out in the eastern Coast Ranges. West of there, jammed against and partially underneath the Coast Range Ophiolite, lie the terrane rocks of the Franciscan Complex, including the chert, pillow basalt, and serpentinite that we explored around the Golden Gate in chapter 1.

Detailed mapping and age dating of California's terrane rocks lets us reconstruct with some confidence how California came into existence. Figure 2.5 illustrates the scene about 120 to 100 million years ago, when the geologic assembly of the state was mostly complete. The figure portrays an east–west slice across the

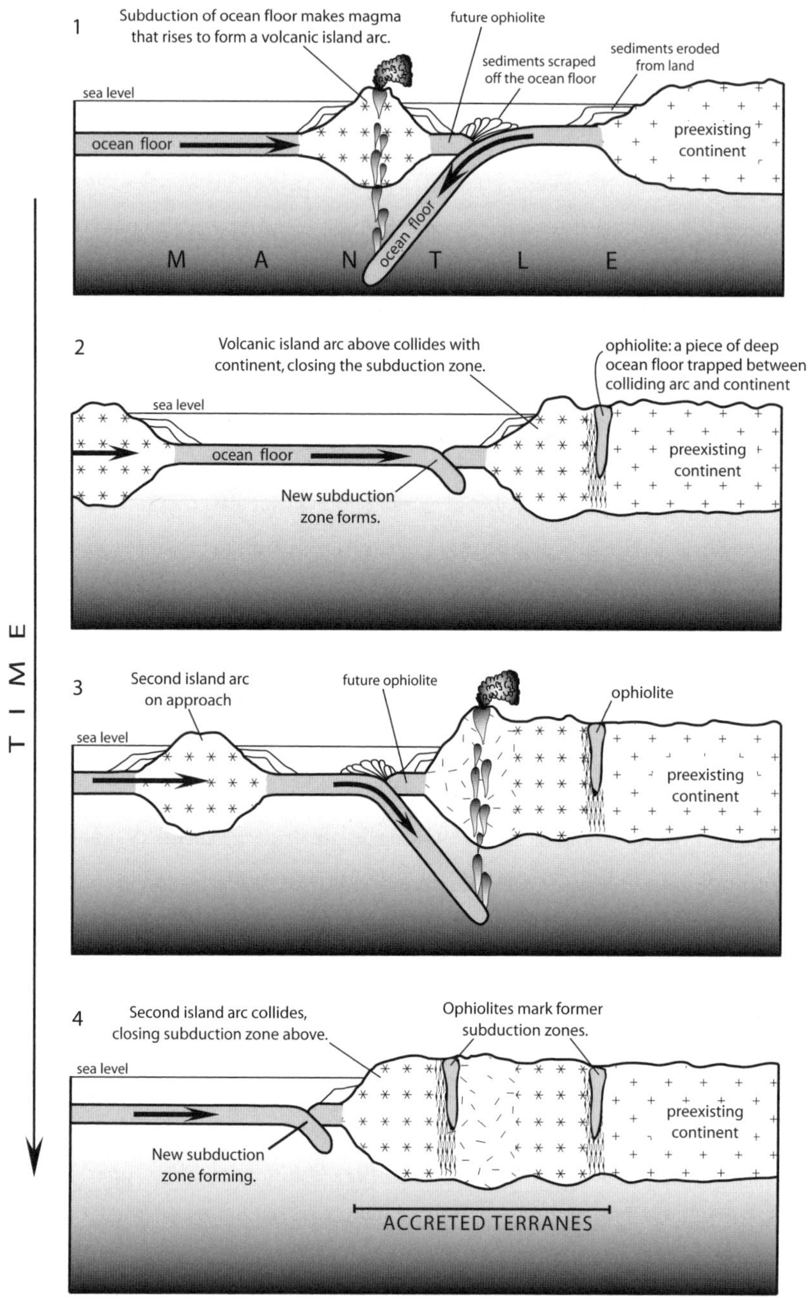

FIGURE 2.2

Continents grow as pieces of oceanic rock collect against their edges. This figure illustrates (very schematically) the growth of California, and by extension, much of the rest of western North America. Over the last 200 million years, subduction has stacked all manner of oceanic real estate against the western edge of North America (see figure 1.4). Collectively, these newly arrived rocks are called accreted terranes. They include ophiolites (pieces of deep ocean floor), volcanic islands and →

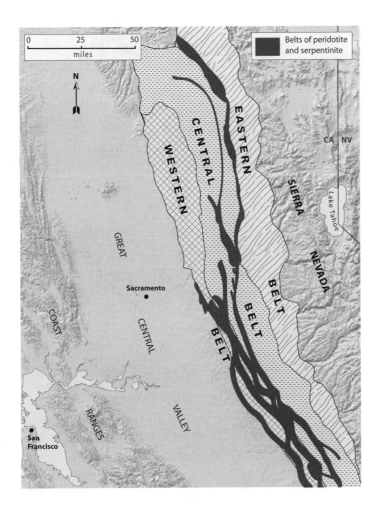

FIGURE 2.3

The accreted terranes of the western Sierra Nevada form three great belts that parallel the geographic trend of the range. The Eastern Belt contains the oldest terranes and the Western Belt the youngest terranes—a consequence of their east-to-west sequence of arrival from the ancient Pacific (see figure 2.4 for an analogy).

Terranes younger than those of the Western Belt underlie the Central Valley, and still younger terranes appear in the Coast Ranges, including the rocks of the Franciscan Complex that we explored in chapter 1. Accreted terranes both assembled California and brought much of the gold that triggered the gold rush.

seamounts, limestone reefs, and sedimentary debris caught up in the jam-ups between colliding bodies of rock. Ophiolites are key indicators of this process because each ophiolite marks the position of an ocean basin that closed to add new rock to the continent's edge. For clarity, the figure greatly exaggerates topographic relief and the depth of the ocean basins.

FIGURE 2.4

A model for the geologic assembly of California, and a tough job for your dedicated author since easy-crush cans typically go with watery beer. Once I finished this diagram, I went back to Sierra Nevada Pale Ale.

FIGURE 2.5 (OPPOSITE)

East–west slice through California illustrating its plate tectonic setting in mid-Cretaceous time, about 120 to 100 million years ago. The left side of the diagram corresponds roughly to where the Bay Area is today, and the right side to what is now the north-central Sierra Nevada. Gold originally arrived in California with the Jurassic ophiolites and other terrane rocks that we see today in the western Sierra Nevada. Much of the gold arrived in deposits of metal sulfides around fossilized seafloor hydrothermal vents. In Cretaceous time, hot underground waters, probably heated by magma rising from the Farallon Plate, remobilized this gold by dissolving it out of the seabed rocks and then re-precipitated it as lodes (gold-bearing quartz veins) along the collision zones of faulted, shattered rock between the Jurassic terranes (bottom diagram). For clarity, the diagrams exaggerate topographic relief.

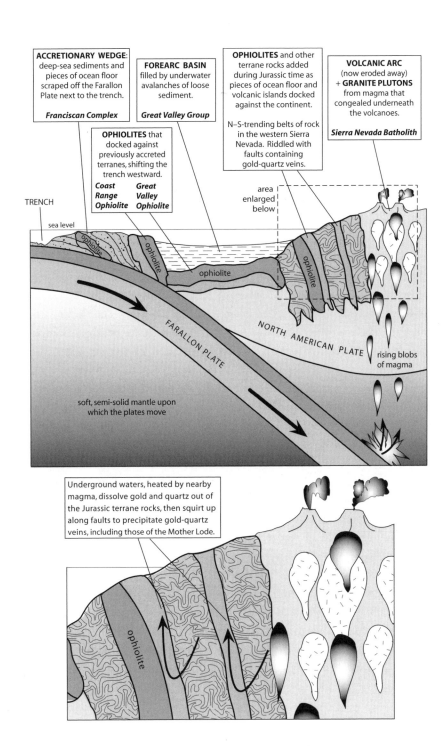

ACCRETIONARY WEDGE: deep-sea sediments and pieces of ocean floor scraped off the Farallon Plate next to the trench.

Franciscan Complex

FOREARC BASIN filled by underwater avalanches of loose sediment.

Great Valley Group

OPHIOLITES that docked against previously accreted terranes, shifting the trench westward.

Coast Range Ophiolite   *Great Valley Ophiolite*

OPHIOLITES and other terrane rocks added during Jurassic time as pieces of ocean floor and volcanic islands docked against the continent.

N–S-trending belts of rock in the western Sierra Nevada. Riddled with faults containing gold-quartz veins.

VOLCANIC ARC (now eroded away) + **GRANITE PLUTONS** from magma that congealed underneath the volcanoes.

Sierra Nevada Batholith

TRENCH

sea level

ophiolite

ophiolite

ophiolite

ophiolite

area enlarged below

FARALLON PLATE

NORTH AMERICAN PLATE

rising blobs of magma

soft, semi-solid mantle upon which the plates move

Underground waters, heated by nearby magma, dissolve gold and quartz out of the Jurassic terrane rocks, then squirt up along faults to precipitate gold-quartz veins, including those of the Mother Lode.

ophiolite

center of the state, roughly between what is now the Bay Area and the Sierra Nevada. On the east (right) side of the diagram, magma rises from the subducting Farallon Plate. This magma will eventually become the granite of the high Sierra Nevada (the story of chapter 4). Immediately to the west lie the three great terrane belts of the western Sierra Nevada (shown in map view in figure 2.3). As this rock crunched into North America during Jurassic time, it advanced the edge of California west to about where the western Sierra Nevada foothills end today. Had you stood 145 million years ago where the Sierra foothills are now, you would have seen no land to the west—no Central Valley, no Coast Ranges. Nothing but the deep blue Pacific Ocean.

Most of the rest of California arrived from the Pacific during Cretaceous time. From about 140 to 130 million years ago, the Great Valley and Coast Range ophiolites crunched into the continental edge. The ophiolites jammed shut the trench (or trenches) that had pulled in the Jurassic terranes, forcing a new trench to form to the west. The dense rock of the Great Valley Ophiolite sagged down, forming a trough called a fore-arc basin west of the Jurassic terranes. Into this trough cascaded countless turbidity currents— undersea avalanches of sand and mud—that stacked up thickly throughout Cretaceous time. Today these rocks—called the Great Valley Group (figure 2.5)—underlie much of the Great Valley and are found tilted up on edge in the eastern Coast Ranges. Meanwhile, the Farallon Plate plunged like a down escalator beneath everything. The Cretaceous edge of the continent (represented by the Coast Range Ophiolite in figure 2.5) scraped like a plow blade against the seafloor, dislodging slivers of abyssal ooze and oceanic crust. The resulting accretionary wedge became the Franciscan Complex, which includes the deep-sea rocks that we explored around the Golden Gate in chapter 1.

If you're feeling a bit overwhelmed by the details at this point, I don't blame you. California's geology is unruly and complicated, and straightening it out has been a monumental task. All honor to the pioneers who set us onto this road—to the geologists who, beginning in the late 1960s, harvested the fresh ideas of plate tectonic theory to lay a new feast on the table of science. They include William Dickinson (who solved the conundrum of California's Cretaceous rocks—the Sierra Nevada Batholith, Great Valley Group, and Franciscan Complex—by fitting them into the tectonic framework shown in figure 2.5), Eldridge Moores (who figured out much of what ophiolites tell us about subduction and the growth of continents), and dozens of others who mapped, sampled, and hammered their way to a better understanding of California's complex birth.

·   ·   ·

Now that we've built California, we can insert lode gold into this picture. Most of California's lode gold—if it isn't in tooth fillings, wedding rings, or bank vaults by now— lies in the Jurassic terrane rocks of the western Sierra Nevada. In other words, the gold, like the rock that built California, is a geologic import, brought here by subduction. The story, as best as we can tell, goes like this:

It is early in Jurassic time, nearly 200 million years ago. North America has torn away from Pangaea. The young North Atlantic forms a widening strait. Ten thousand miles to the west, at a now-vanished mid-ocean ridge in the ancient Pacific, the seabed is splitting and growing. Seawater seeps down through cracks in the hot rock. Percolating down perhaps a mile or more, it heats four to five times hotter than its surface boiling point, but it doesn't boil; the pressure won't let it. Loaded with chlorine atoms from sea salt, the super-heated water becomes intensely corrosive. It percolates through the rocks like venomous brine, dissolving atoms of gold, copper, zinc, lead, and other metals from the surrounding crust. The metal-saturated waters then rise through crevices and jet out from the seabed as scalding underwater geysers. Where hot water hits cold, the dissolved metals fall out of solution, forming deposits of metal-rich sulfide minerals. The sulfides pile up into grotesque, black chimneys around these hydrothermal vents.[3] Gold atoms find themselves jailed within the molecular lattices of the precipitating sulfide minerals.

The next step is to send this trapped gold to California—or, more exactly, to where California is being assembled. Fast-forward now to later in Jurassic time, about 160 million years ago. The seafloor with its gold cargo collides with the then-western edge of North America, stuffing shut the trench that reeled it in. Today these rocks constitute most of the Western Belt of the Sierra Nevada foothills (figure 2.3), including the Smartville Ophiolite (figure 2.1). The newly arrived rocks jam against those of the Central Belt in a slow-motion smashup, shattering the rock along their mutual glue-line. That glue-line will come to house the Mother Lode—California's renowned lode gold belt. But there's no gold there yet. It's still trapped around the fossilized seafloor hot vents.

Fast-forward now to the middle of Cretaceous time, about 125 to 110 million years ago. More seafloor rocks have docked west of the Jurassic terranes, displacing the subduction trench westward to the position shown in figure 2.5. East of the trench, magma ascends in viscous blobs from the Farallon Plate, rising to solidify into granite in the upper reaches of the crust. The rising magma heats up the Jurassic terrane rocks and their resident gold. The gold atoms, in a case of chemical déjà vu, find themselves kidnapped once again by hydrothermal waters—but with a difference. This time, along with gold, the hot waters gather up vast amounts of silica—a chemical compound abundant in granite and other rocks of the continental crust. Saturated with gold and silica in solution, the hot waters rise buoyantly, seeking out the easiest pathways—the glue lines of weak and shattered rock between the once-separate Jurassic terranes (bottom diagram in figure 2.5). There, within the shattered rock, the waters drop their loads of silica as veins of quartz, milk-white and harder than steel.[4] The gold atoms drop out too, as flecks (sometimes visible, but usually microscopic) entombed in the quartz. Thus is born a lode—a gold-bearing quartz vein (figure 2.6). Gold rush miners knew that where you find quartz veins, you often find granite nearby. "Quartz and granite appear to hold companionship," Alonzo Delano observed in 1849—although he didn't

FIGURE 2.6

A typical lode, or quartz vein, in the western Sierra Nevada stands out light in color against the darker terrane rock into which it has intruded. "Through the whole chain of the Sierra Nevada," forty-niner Alonzo Delano observed, "innumerable veins of quartz extend, bearing a general parallel with the principle chain. They vary in thickness from a few inches to fourteen feet [and] are found dipping in various directions, but not infrequently are perpendicular." The quarter coin near the bottom of the picture gives scale.

know why. Hydrothermal waters, we now believe, form the link between granite, quartz, and gold.

Dense concentrations of lodes define lode districts. And no lode district stands taller in California's history than the Mother Lode district. Comparing figures 2.3 and 2.7, you can see that the Mother Lode parallels the belts of Jurassic terrane rocks. But you don't need a geologic map to find the Mother Lode. Any state highway map will do. Just find Highway 49—the gold rush highway—which follows the Mother Lode for its entire 120 miles from Mariposa to Georgetown. Quartz veins shine from the road cuts, and the roadside markers for the now-dead gold mines read like the pages of gold rush history: Eureka, Kennedy, Carson Hill, Argonaut, Keystone, Utica, Eagle-Shawmut, Georgia-Slide, Rawhide. "The mere thought of this huge body of ore lying beneath the

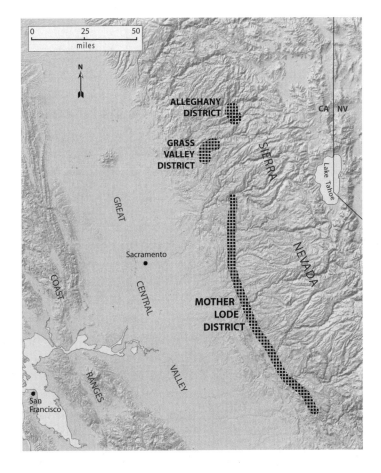

FIGURE 2.7

Most of California's lode gold came from three mining districts in the Sierra Nevada: the Mother Lode, Grass Valley, and Alleghany districts. The Mother Lode district forms a 120-mile-long belt that follows the glue-line between Jurassic terranes—a connection readily apparent if you compare this map with figure 2.3. In Cretaceous time, hydrothermal waters invaded the shattered rock of the terrane collision zone to deposit dense networks of lodes. Although the Mother Lode district had by far the greatest number of mines, California's most productive lode mine, the Empire Mine, was in the Grass Valley district.

[Sierra Nevada] foothills set the crustiest miner quivering," historian H. W. Brands writes of the Mother Lode. "In some cases, awe induces respect. In this case, it elicited avarice, as the miners began to plot how to rip the gold from the womb of its mother."

Ripping gold from Mother Lode quartz veins would not prove easy. Lode mining (also called quartz mining or hard-rock mining) was capital-intensive. It was, therefore, usually a company effort, funded by investors, in which miners worked for wages, not gold. This is unlike placer mining—at least in its early years—when individuals or small

FIGURE 2.8

A ten-stamp mill at the Empire Mine near Grass Valley, with my very good dog Scout for scale. The large disk on the left side of the stamp mill held a steam-driven pulley. As the pulley spun the shaft, the cams raised and then dropped the heavy stamps onto blocks of quartz. With a roar louder than any freeway today, the stamps pulverized the quartz into powder, freeing the microscopic flecks of gold.

groups worked their own streambed claims and counted on personal initiative and luck to win their fortunes—or at least their dinner. Lode mine investors were often distant financiers: soft, pink-handed men who earned their livings with pens, not hammers. Whereas placer (that is, riverbed) mining crested and crashed within a decade of 1849 (see figure 3.3), lode mining ramped up gradually and did not enjoy wide profitability until the 1860s. In effect, lode mining was the tortoise to placer mining's hare: slower to start, but longer to last.

The first step in mining lode gold was to bore tunnels or shafts along sets of quartz veins, like termites following wood grain. Miners drilled holes in concentric circles in the vein rock. They filled the holes with black powder (or, later, dynamite, which Alfred Nobel patented in 1867), and then set the charges with fuses of varying lengths to create a timed blast—center holes first, outer holes last. This sent the exploded rock in toward the growing center of the blast hole, creating a controlled collapse that maintained the shape of the tunnel. They then hauled out the blasted rock using skiffs—small carts that ran on miniature railroad tracks. In early lode mining days, teams of donkeys pulled the skiffs through the narrow tunnels. These unwilling troglodytes rarely saw the light of day after they were lowered (using great slings around their bellies) down the long vertical elevator shafts to the underground tunnel networks. Entire populations of donkeys lived and died in California lode mines, for it was far cheaper to bring hay in, and haul manure out, than it was to bring the animals up to the surface. In later years, steam power and then electric power hauled the skiffs with cables and pulleys.

Once the blasted rock emerged from the mine, miners set about separating chunks of quartz ore from waste rock. They sledge-hammered the quartz down to fist-sized pieces, which then went to a processing mill. There, batteries of steam-powered steel stamps smashed the quartz to powder (figure 2.8). When Mother Lode mining was in full swing, it was said that no place in the western Sierra Nevada was out of earshot of these thundering stamp mills. The powdered quartz was then mixed with water, and the resulting slurry sent to concentrating stations where the microscopic gold was separated. The earliest method of separation sent the slurry through a sluice box—a long wooden trough lined with riffles (short sticks set like speed bumps across the flow). Since gold is heavier than anything it travels with, the gold particles settled toward the bottom of the flowing slurry and collected against the riffles. Miners learned that they could catch more gold by coating the riffles with mercury, with which gold readily amalgamates. Later methods flipped the recovery process on its head by floating the gold out of the slurry rather than taking it off the bottom. Vats of slurry were injected with bubbled air and gold-attracting chemicals. The microscopic gold flecks were drawn chemically to the air bubbles and floated off the top.

The one constant in a lode miner's life was water. It dripped and flowed incessantly in the mines, where it rotted supporting timbers, flooded low places, and generally made life miserable. Pumps chugged without letup, for if ever they stopped, the tunnels and shafts would quickly flood. A correspondent for *Hutchings' California Magazine* in 1857 described his visit to the soggy guts of a lode mine:

Drip, drip, fell the water, not singly, but in clusters of drops and small streams. . . . The miners, who were removing the quartz from the ledge, looked more like half-drowned sea lions than men. We did not make ourselves inquisitive enough to ask the amount of wages they received, but we came to the conclusion that they must certainly earn whatever they obtained. Stooping, or rather half-lying down upon the wet rock, among fragments of quartz and props of wood, and streams of water, with pick in hand, and by a dim but waterproof lantern, giving out a very dim and watery light, just about bright enough, or rather dim enough, and watery enough, as Milton expresses it, "to make darkness visible," a man was at work, picking down the rock—the gold-bearing rock.

A placer miner's life also centered on water. But for him, water was not a nuisance. It was, instead, a weapon—a hydraulic weapon, to be used with devastating force against fossil riverbeds of gold, as we'll see in the next chapter.

NOTES

1. To estimate what gold rush–era dollars were worth in today's dollars, multiply by twenty-six. From historical currency conversion tables by Robert C. Sahr, Political Science Department, Oregon State University, Corvallis.

2. Figuring out when individual terranes arrived—or "docked" against a continent—is tricky. The best hints come where bodies of magma have cut through multiple sets of terrane rocks. In such cases, cross-cutting logic tells us that the terranes docked before the magma intruded. The intruding rocks (which in the Sierra Nevada are usually granite plutons) can be age dated using radioactive decay. (See appendix I for an explanation.) The ages I give in this chapter are based on this approach, as summarized in Dickinson 2008.

Terrane experts generally concur that the terrane rocks of the western Sierra Nevada group into three broad belts with the oldest rocks to the east and the youngest to the west. But little consensus exists on the details. There are several open questions: How many individual terranes exist within these belts? How far did they travel before docking on North America? Which terranes arrived separately versus already glued together? How many subduction zones closed and which way were those subduction zones facing, to account for the geologic relationships that we see? For more, see Moores et al. 2006 and Dickinson 2008.

3. Oceanographers see this process—called hydrothermal precipitation—happening today around hot vents at mid-ocean ridges. They call the vents "black smokers" for the billowing dark plumes of metal sulfides that precipitate from the scalding waters. The chimneys that grow around these vents typically range from a few feet to a few tens of feet high, although one monster chimney on the Juan de Fuca Ridge (known as Godzilla) is as tall as a skyscraper.

4. A vein is a deposit of minerals that fills in a fracture or a fault. Inject some glue into a deep crack and let it solidify; that mimics how magma or hydrothermal fluids make a vein. Unlike the veins in your body, a geologic vein has a tabular or sheet-like shape.

3

RIVERS OF GOLD

"Civilization exists by geological consent," the historian Will Durant once remarked. He was right. Imagine civilization without geologic consent. You'd have no electricity. (We get most of our juice from burning coal, the remains of fossilized plants.) You'd have no fuel or lubricating oil for your vehicle. (They come from crude oil.) But that wouldn't matter because there would be no roads to drive on. (Asphalt comes from crude oil; concrete from limestone and gypsum.) And *that* wouldn't matter because there would be no planes, trains, or automobiles. (Steel comes from iron ore.) You'd have no clothing that you didn't make yourself, no food that you didn't grow or kill for yourself. (Goods move through our economy via fossil fuels.) If you can identify a single day in your life when you didn't depend on geologic consent, I'll buy you a beer. It will be a cold beer (refrigerated by coal-fired electricity) of your choice (thanks to fossil fuel-based transportation) in either an aluminum can (from bauxite ore) or a glass bottle (from quartz sand), paid for by my salary as a geology teacher (money from rocks) and served with a grin corrected by porcelain (from kaolinite—a clay mineral).

Durant was right, but Durant's saying also works in reverse. Sometimes geology exists by civilization's consent—or ceases to exist. In the western Sierra Nevada, forty billion cubic feet of the mountains have ceased to exist thanks to nineteenth-century miners in pursuit of gold.

The jet pulls out of the fog soup that hovers over San Francisco Bay and banks east toward Reno. The wrinkles of the Coast Ranges fade to flat as you pass east over the Central Valley, hazed with agricultural smog. East of Sacramento, the western Sierra

west-tilted upland surface of the Sierra Nevada
(slope ~180 feet per mile)

WEST

EAST

OMEGA DIGGINS

FIGURE 3.1

View north from an overlook along Highway 20 toward the valley of the South Yuba River showing the scars of the Omega Diggins hydraulic mine in the middle distance. The excavation spreads one mile long by a half-mile wide. From the mid-1850s until 1884, miners targeted gold-rich fossil riverbeds at this and dozens of other hydraulic mines throughout the western Sierra Nevada. The slope of the mountain's upland surface reflects the westward tilt of the Sierra Nevada as it rose, as shown in figure 3.5.

Nevada foothills begin, rising forest-green from the dust-brown valley floor. You can almost smell the change as your eye follows the slopes up into thinner, cleaner air.

That's when you see the scars. They appear as misshapen splotches of roseate white, several miles across, like splashes of milk on a carpet of green. Most of the splotches lie on the ridge tops between the deep canyons that crease the Sierra Nevada's western slope. Something once hacked at the Earth's face here, and the coniferous forests—slow to retake bare rock—have only partly healed the wounds.

Highway 20 between Grass Valley and Emigrant Gap gives you a closer view of one of these scars. It's called Omega Diggins (figure 3.1). Nineteenth-century miners blasted enough gold-bearing gravel from the mountain here to fill twenty-five football stadiums. Their weapon was not black powder or nitroglycerine, but water—shot with terrifying force from gigantic hydraulic cannons.

Omega Diggins is the remains of a modest-sized hydraulic gold mine. Five others occur along the South Yuba River, and several dozen more lie scattered throughout the drainage basins of the Feather, Yuba, Bear, and American rivers. All were gouged out during three frenetic decades from 1853 to 1884. The largest is Malakoff Diggins, along

FIGURE 3.2

Hydraulic mining at Malakoff Diggins in about 1876 (top), along with a close-up view of a monitor at Malakoff Diggins State Park today (bottom). (Top photograph from the Hearst Collection of Mining Views by Carleton E. Watkins, Bancroft Library, University of California, Berkeley.)

the South Yuba River about ten miles west of Omega Diggins. The Malakoff hydraulic pit measures 7,000 feet long, 3,000 feet wide, and up to 600 feet deep, all blasted out by high-pressure water. You can wander the devastation today at Malakoff Diggins State Park—probably the only park in the world dedicated to a man-made hole in the ground (figure 3.2).

All told, hydraulic miners stripped forty billion cubic feet of gold-bearing gravel out of the Sierra Nevada.[1] That's five times more earth than was moved during the excavation of the Panama Canal, which is widely (if incorrectly) cited as the greatest earth-moving operation in history up to that time. Although "hydraulicking" ended in the Sierra Nevada more than 120 years ago, it bequeathed a toxic legacy down to present generations. Hydraulic miners poured 2.3 million of gallons of mercury into their sluices to help bind up the gold. Some 70,000 to 80,000 gallons of it leaked out.[2] Today it contaminates fish and groundwater throughout the western Sierra Nevada. But a bigger story lies in the turbid slurries of waste gravel—a mountain of dirt that triggered California's first man-made ecological disaster.

.   .   .

The story of hydraulic mining begins in 1849, with the gold rush. From all corners of the Earth, men (and a few women) converged on the western Sierra Nevada to dig for placer gold—flakes and nuggets of gold eroded out of bedrock and washed downhill into riverbeds. Gold is nearly twice as dense as lead, so even small pieces work their way down through riverbed gravels to collect on channel bottoms. You have to sift a lot of gravel to get to this pay dirt. The imperative to sift more and more gravel forced miners to adopt new technologies. Solitary panning gave way to teams of miners who worked rockers, long toms, and sluices. Sluices grew larger, segment by segment like tapeworms, to handle more gravel. Larger sluices demanded more water, which meant dams, ditches, and flumes—sprawling and interconnecting in vast confusions up and down the river valleys of the western Sierra Nevada.

And the gold poured forth. It came out of the mountains in astonishing amounts—sums the world had never before seen: 17 tons of pure gold in 1849, 68 tons in 1850, 126 tons in 1851, and 135 tons in 1852. Then, almost as fast as it had soared, production began to plummet. Less than ten years after the gold rush began, the riverbeds of the western Sierra Nevada had been practically wrung free of gold (figure 3.3).

Even before riverbed gold production began to peter out, miners began searching uphill for the source of all that riverbed gold. Crawling over the ridges between river valleys in the western Sierra Nevada, they soon realized that a lot of the gold was coming out of *fossil* riverbeds, now high and dry on the valley walls and ridges. These ancient riverbeds were gigantic, like fossilized Colorado Rivers, with channels several hundred feet deep in gravel. The miners dubbed them the auriferous gravels (*auri* = gold, *ferous* = to carry or bear).

The challenge of mining the auriferous gravels was that the fossil riverbeds lay high on sun-baked ridges hundreds of feet above existing streams. The miner's solution was to build wood-and-stone dams far upstream and send the impounded water through miles of ditches and flumes to points *above* the fossil riverbeds. From there, the miners directed the water steeply downhill through riveted steel pipes and canvas hoses to monitors—huge, swivel-mounted water cannons. Backed by immense pressure, the

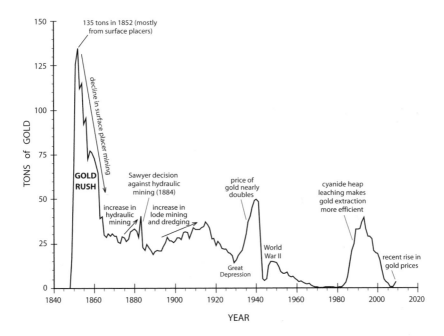

FIGURE 3.3

California gold production from 1848 to today. The gold rush peak was mostly from modern riverbeds (surface placers). Post–gold rush production included hydraulic mining of fossil riverbeds (until the practice was banned 1884), lode mining (extraction of gold from quartz veins—the story told in chapter 2), and large-scale dredging of riverbeds from barges.

water roared from the monitors at one hundred miles per hour in columns as thick as a telephone pole. The sprays arced hundreds of feet through the air and crashed with devastating force against the auriferous hillsides, turning them into brown torrents of clattering rock (figure 3.2). The turbid slurries were funneled through sluices where the heavy gold particles collected against riffles, aided by large doses of mercury.

Hydraulic mining was probably rather inefficient in the fraction of gold that it recovered. But hydraulic miners could blast apart so much riverbed gravel so quickly that the method netted huge sums of gold. By the 1870s, hydraulicking had temporarily reversed the long post–gold rush decline in California's gold production (figure 3.3).

The success of hydraulic mining also wrote its death sentence. Even a modest-sized hydraulic mine used more water each day than the growing city of San Francisco. The gravel slurries cascading from the auriferous hillsides poured into the beds of the Feather, Yuba, Bear, and American rivers. Displaced by the rising tide of rock debris in their channels, the rivers erupted across the landscape in wave after wave of turbid floods, laying waste to thousands of acres of farmland across the Central Valley. The citizens of Marysville threw up levees to protect their town from the swollen Yuba River. The levees broke, filling Marysville with four feet of soupy mud. The debris surged on to the lower Feather and Sacramento rivers, where it piled up in massive sand bars that ran steamboats aground. Mud and silt smothered the deltas at the head of San Francisco Bay and stained the bay brown all the way to the Golden Gate. A reporter for the *San Francisco Daily Evening Bulletin* described the flood-wrought devastation across the Central Valley in March 1878. In all directions lay "a wild waste of waters, broken levees, rows of earth bags, fences peeping out of the water, and houses in the water. . . . The beautiful orchards were either in the water or covered with slime. . . . The population along the [Sacramento] river fought the floods as men never fought anything else since the world began, but all in vain."

The deep pockets of the hydraulic mining companies kept the legal wolves at bay for several decades as profits soared. But in 1884, the Anti-Debris Association—a group of farmers, ranchers, and townspeople that rallied to battle the hydraulic miners—finally won their case in federal court. The court's decision banned the flushing of mining effluvia into the tributaries of the Sacramento and San Joaquin rivers. The ruling, by federal judge Lorenzo Sawyer, did not outlaw hydraulic mining per se; miners could still hose down mountainsides as long as they trapped the slurry behind debris dams. But the cost of debris management proved to exceed profits, so the ruling effectively ended hydraulic mining in the Sierra Nevada from 1884 onward. (Limited hydraulicking continued in the Klamath-Trinity region of northern California until the 1950s.)

.   .   .

How did fossil riverbeds pregnant with gold end up high and dry in the Sierra Nevada? A map of the fossil riverbeds (figure 3.4) tells part of the story. In all rivers, tributaries converge on the main channel downstream like roots on a tree. Figure 3.4 thus shows that the fossil rivers flowed generally west. The gold they contain must therefore have eroded out of mountains to the east, where Nevada is now. Fossil plants preserved in the riverbeds show that the rivers flowed about forty-five to forty-eight million years ago, during Eocene time.

Today, no river that originates in Nevada can flow west across California. The eastern wall of the Sierra Nevada turns them all back to die in the Great Basin—a huge topographic bowl, centered on Nevada, where rivers flow in, never out. But the Great Basin is a recent development in the geologic story of the West. Back in the days when the fossil rivers flowed, Nevada was a lot higher than it is today, and rivers ran west all the way from there to the Pacific Ocean. Elevate Nevada into rugged plateau more than two miles high, much like the Altiplano of the central Andes today, and you have an image of the *Nevadaplano*—a highland region that dominated Nevada before it collapsed like a punctured soufflé to make the Great Basin (a story that we'll explore in detail in chapter 6). Visualize snowfields draped between the rugged peaks. Meltwaters seep across the landscape. Flakes and nuggets of gold crumble from the bedrock. Rain and meltwater nudge those bits of gold downhill toward stream channels. Small channels merge into bigger channels, which merge into foaming white-water rivers as big

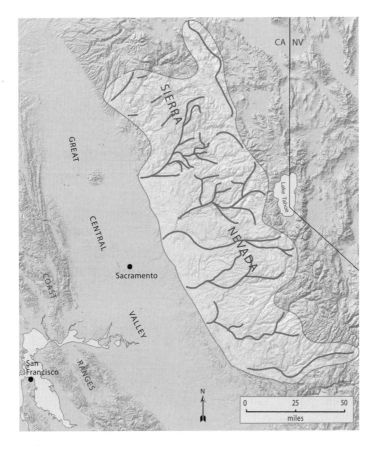

FIGURE 3.4

Eocene (roughly forty-five-million-year-old) fossil riverbeds of the Sierra Nevada. The branching pattern of the tributaries shows that the rivers flowed west out of the Nevadaplano, a highland region that once existed where Nevada is today.

as the Colorado. Softball- and basketball-sized boulders roll west along the channels toward California—along with plentiful gold.

The Sierra Nevada today lies about where the western flank of the Nevadaplano once sloped toward the Pacific Ocean. In effect, the Sierra is what remains of the Nevadaplano's western edge. The rest of the Nevadaplano was destroyed by the stretching and collapse of the crust that made the Great Basin (figure 3.5).

What evidence supports this story? The first is fossil plants indicating that Nevada once stood about one mile higher than it does today—high enough to have been a source for the fossil riverbeds. The second is evidence from ancient raindrops preserved within the fossil riverbeds. Let me explain both.

Anyone going up or down a mountain can see how the foliage changes with elevation. Plants that are adapted to warmer, drier environments lower on the mountain

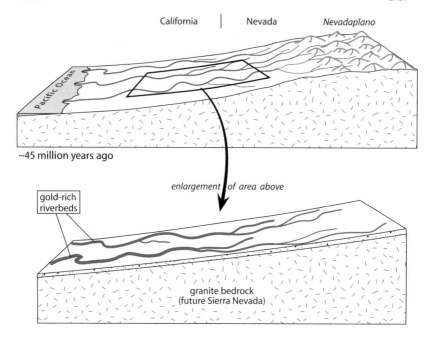

WEST — EAST

California | Nevada   *Nevadaplano*

Pacific Ocean

~45 million years ago

enlargement of area above

gold-rich
riverbeds

granite bedrock
(future Sierra Nevada)

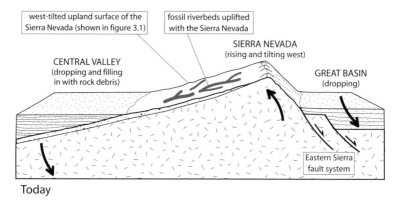

west-tilted upland surface of the
Sierra Nevada (shown in figure 3.1)

fossil riverbeds uplifted
with the Sierra Nevada

SIERRA NEVADA
(rising and tilting west)

CENTRAL VALLEY
(dropping and filling
in with rock debris)

GREAT BASIN
(dropping)

Eastern Sierra
fault system

Today

FIGURE 3.5

Forty-five million years ago, large rivers tumbled west from the Nevadaplano—a gold-rich highland plateau that once lay where Nevada is now—toward the Pacific Ocean (top). By about sixteen million years ago, east–west stretching of the crust had triggered the gradual collapse of the Nevadaplano to form the Great Basin (bottom)—a process that continues today (see chapter 6). The Sierra Nevada represents land that stayed high along the Nevadaplano's western edge. Starting about five million years ago, the Sierra Nevada gained several thousand more feet of elevation as it tilted west like the rising end of a seesaw. West-flowing streams sliced into the mountain as it rose, cutting down through the fossil riverbeds to carve the deep, steep-walled canyons that now bisect the Sierra's western slope.

give way to plants that thrive in cooler, wetter conditions higher up. Going from the Anza-Borrego Desert up into the mountains east of my San Diego home, I know I've hit 3,200 feet when juniper trees appear, and 4,500 feet when pines take over from chaparral. Within a given climatic region, plants can serve as proxies for elevation. Fossil plants from western Nevada indicate that, up until about sixteen million years ago, Nevada lay about two miles above sea level. The difference between that and today's one-mile-high average elevation reflects the down-dropping of the Nevadaplano to form the Great Basin.

Plants aren't the only thing to change as you go up or down a mountain. The nature of raindrops changes as well. Ancient raindrops contain a chemical signal that can tell us the elevation at which the rain fell. Here's the idea: the hydrogen in water molecules (H_2O) comes in one of two forms, or isotopes: a lighter form (protium) and a heavier form (deuterium). The difference means that individual water molecules differ slightly in weight depending on whether they contain heavy or light hydrogen. As moist wind rises up a mountainside, the air expands and cools, dropping its moisture as rain or snow. Heavier water molecules tend to fall first, at lower elevations. As the air continues up the mountain, the clouds are progressively depleted of heavier water and drop a greater fraction of lighter water. You can quantify this change by catching rain at various elevations on a mountain and measuring the proportion of light versus heavy water molecules. If you then compare what you find in present-day raindrops to *ancient* raindrops, you can tell roughly the elevation at which the ancient raindrops fell. But how to catch ancient raindrops? Soil, it turns out, can catch and hold water for millions of years by chemically binding water molecules onto clay particles. Therefore, if we extract ancient rainwater from old soil beds and measure the fraction of light versus heavy water, we can get a good idea of the elevation at that location at the time the rain fell.

Geologists at Stanford University have done this by collecting ancient rainwater trapped in old soil layers preserved close to the fossil riverbeds of the Sierra Nevada. (The fossil riverbeds still lie widely exposed in old hydraulic pits throughout the Sierra Nevada, so sampling them is quite straightforward.) The geologists collected ancient soil samples across a range of elevations and compared the amount of light versus heavy water to modern rainfall at the same elevations. They found little difference between the samples of present-day rainwater and ancient rainwater. Conclusion: the fossil riverbeds today lie at pretty much the same elevation as they did when they flowed forty-five million years ago.

This finding has knocked the socks off more than a few geologists. Before the Stanford study, most of us had assumed that the land that now forms the Sierra Nevada lay close to sea level before rising to its current lofty heights. The raindrop data suggest a new way to think about the Sierra Nevada. The range forms high, bold topography today not so much because it has risen *up* from a low elevation, but because the even higher land that once lay to the east—the Nevadaplano, the source of the auriferous gravels—has dropped *down* (figure 3.5).

In addition, during the last five million or so years, the Sierra Nevada has added several thousand feet to its elevation by tilting west like the high end of a seesaw. The Sierra Nevada and the Central Valley together form a single, west-tilting block of the Earth's crust. The up-tilted part of the block forms the mountains; the down-tilted part forms the valley (figure 3.5). From almost any high place on the Sierra's western slope—such as the Omega Diggins overlook, shown in figure 3.1—you can see this westward tilt. (Compare the bottom diagram in figure 3.5 to the photograph in figure 3.1 to see what I mean.) As the mountain tilted west, west-flowing rivers responded to the steeper slopes by slicing downward, cutting the deep, sharp valleys that furrow the range's western slope today. The rivers shipped trillions of tons of eroded rock debris out to the Central Valley, where the weight (shifted now from the mountain to the valley) kept the seesaw tilting. Wherever the rivers cut into the gold-bearing fossil riverbeds, they recycled the gold into their own channels—gold that contributed to the great rush of 1849. Wherever erosion spared the fossil riverbeds, they remained undisturbed for a few more years—until miners began their hydraulic assaults during the giddy days of gold.

· · ·

In the up-and-down country between the middle and north forks of the Yuba River lies a branch of one of the fossil riverbeds. When gold rush miners traced its sinuous channel, they found that it went under the tiny town of Camptonville. A gaping hydraulic valley soon filled the space where old Camptonville used to be. A new Camptonville rose nearby, next to the town cemetery, at a safe distance from the auriferous channel. When I say that the Camptonville Cemetery is a place that's hard to leave, I'm not attempting morbid humor. It sits on a breezy knoll studded with stately oaks and has the prettiest view in a region of pretty views, taking in acres of rolling Sierra Nevada foothills cloaked with cedar, oak, madrone, and pine.

Jonathan Spencer took up residence in the Camptonville Cemetery in September 1853. Like many gold rush–era men, it's hard to know much about him. Few miners kept personal records. He was probably one of thousands who scrabbled for placer gold along the banks of the Yuba River. What sort of man was he? Did he crack jokes, laugh loudly, and grab hold of life with gusto? I hope so. His son must have loved him, or at least honored him, for when Warwick Spencer buried my great, great, great grandfather, he gave him a living legacy. The son planted four seedling incense cedars at the corners of his father's grave. Today these trees—now more than a century-and-a-half old—soar one hundred feet above the gravesite. Planted too close for the size they would attain, the trees now lean slightly away from one another, so that the space between them opens toward the heavens. The trees almost certainly contain a fraction of my gold rush ancestor, for life's carbon is endlessly recycled, and Mr. Spencer was just borrowing those atoms for a while, as we all do. His remains, reconstituted, rustle in the limbs above my head, casting a comfortable shade.

NOTES

1. Volume of hydraulic gravel displaced is from Alpers et al. 2005. I estimate the volume of the Omega Diggins hydraulic mine at 1.1 billion cubic feet and the Malakoff Diggins mine at 4.2 billion cubic feet, based on dimensions from topographic maps.

2. Nearly all of the mercury used in California gold mines came from mercury mines in California's Coast Ranges, primarily from red cinnabar minerals produced by hydrothermal alteration of serpentinite within the Franciscan Complex. These mercury mines were so productive that more than three-quarters of the mercury was not used in California but instead exported to nations around the Pacific Rim, or to mines in Nevada and other parts of the West.

4

A TRAVERSE ACROSS THE
RANGE OF LIGHT

In the spring of 1868, a broke, grubby drifter arrived by steamer into San Francisco Bay. City ways did not set well with him, and he soon inquired about the quickest way out of town. "'Where do you want to go?' asked the man to whom I had applied. 'To any place that is wild,' I said.'"

The man directed the wanderer to the Oakland ferry, and on April 1, 1868, he set out on foot for the Sierra Nevada. "It was the bloom time of year," he remembered, and "the landscapes of the Santa Clara Valley were fairly drenched with sunshine . . . the hills so covered with flowers that they seemed to be painted." He climbed to the top of Pacheco Pass, in the Coast Ranges, and from there looked east upon a scene that would change his life—and California's future:

> At my feet lay the Great Central Valley of California, level and flowery, like a lake of pure sunshine, forty or fifty miles wide, five hundred miles long, one rich furred garden of yellow *Compositoe*. And from the eastern boundary of this vast golden flower bed rose the mighty Sierra, miles in height, and so gloriously colored and so radiant, it seemed not clothed with light, but wholly composed of it, like the wall of some celestial city.

The U.S. environmental movement arguably began at that moment. But John Muir always felt that the Sierra Nevada was misnamed.

After ten years of wandering and wondering in the heart of it, rejoicing in its glorious floods of light, the white beams of the morning streaming through the passes, the noonday radiance on the crystal rocks, the flush of the alpenglow, and the irised spray of countless waterfalls, it still seems above all others the Range of Light.

California owes a debt to John Muir. Thanks largely to Muir, thousands of square miles of the Sierra Nevada that might otherwise have been logged or mined out of existence are protected by law. But as much as I admire Muir and what he worked for, I lose patience with his plodding nineteenth-century pace, pausing for rapture around every bend in the trail. "Look at the glory! Look at the glory!" Okay, but let's get on with it. It's time to set the cruise control and bag Muir's range in time for lunch. The Sierra Nevada has a lot to offer any geology buff, but I've got one goal right now: to traverse the range along Interstate 80 and tell the story of its granite.

Accelerating out of Sacramento, I first cross miles of flat suburbs. The dark woods of the Sierra Nevada fill the horizon ahead, but the range doesn't begin until twenty-five miles east of the capital, where the highway rises like a jet in takeoff to climb into the foothills. In the 1860s, the location of the Sierra Nevada's western edge was of keen interest to the Central Pacific Railroad (the western arm of the transcontinental railroad). The federal government had agreed to fund the railroad with construction loans at rates adjusted to terrain: $16,000 per mile over flat country, $48,000 per mile over mountains. Upon learning this, Josiah Whitney—the state geologist of California, and a friend of the railroad—made an unexpected discovery. The Sierra Nevada, he found, begins only seven miles east of Sacramento. I challenge you to go seven miles east of Sacramento and figure out which way is uphill. But that didn't concern Whitney. He knew that the seven-mile point is where the flat layers of sand and gravel that fill the Central Valley transition subtly into somewhat older layers that, because of the Sierra Nevada's rise, tilt a wee bit toward the Pacific. That was enough for Whitney to define the edge of the mountains. Thanks to him the Central Pacific Company collected government loans at the higher mountain rate for laying rails over miles of near-flat terrain.

Climbing into the Sierra Nevada's oak-studded foothills, I whiz past road cuts shining with dark, tectonically smashed-up rock. I've entered the western foothill terranes—those great belts of imported oceanic rock that arrived in Jurassic time, carrying gold from the far reaches of the ancient Pacific (see figure 2.3). At Auburn, the interstate crosses the Mother Lode belt, and history buffs exit here onto Highway 49—the gold rush highway—to explore the historic towns and shuttered mines where men once hacked at veins of quartz. I continue east with the interstate, climbing higher. Near Clipper Gap, I slow and weave at the sight of road cuts bulging sensuously with pillow basalts—the remains of the Lake Combie Ophiolite, a chunk of abyssal ocean floor marooned against the edge of North America during Jurassic time. "At work or play, a geologist always drives like Egyptian painting—eyes to the side," John McPhee once observed. Don't drive with a geologist if you can help it, at least not west of the Rockies.

Climbing higher, I gain the ridge top between the canyons of the Bear River to the north and the North Fork of the American River to the south. On the ridge, the highway rises above the terrane rocks and comes into the auriferous gravels—those gold-bearing river gravels washed westward by rivers from the high Nevadaplano during Eocene time. Here the road cuts glow red and orange from a penetrating rust developed as the riverine landscape soaked under the dripping climates of the Eocene Period. Between Gold Run and Dutch Flat, the interstate swings past the headwall of a massive hydraulic pit. The scoop-like fossil river channels stand out clearly, filled with cobbles as big as baseballs. I slow and weave while other motorists honk in annoyance, but continue on. It's the granite I'm after—up ahead.

Farther on, the road cuts change again as I pass through the layered remains of volcanic ash and lava. Erupted out of volcanoes in Nevada during Oligocene and Miocene time, the ash and lava buried the auriferous gravels—and everything else in the Sierra Nevada—under layers of crackling-hot rock. Later, as the Sierra Nevada rose and tilted west (figure 3.5), west-flowing rivers sliced through these volcanic layers and on down into the rocks beneath to carve out the Sierra's great western canyons. The volcanic layers survive today on the ridge tops between the deeply eroded canyons.

East of Emigrant Gap, the interstate drops off the ridge and descends again into the dark-colored Jurassic terrane rocks. Then, where the highway sweeps left along the Yuba River at Big Bend, the landscape shifts suddenly from dark to light.

This is it!

The color shift marks a profound geologic boundary: the contact between dark terrane rock to the west and the gray-white granite that dominates the high Sierra Nevada to the east. Off the interstate now, I park and set out along a foot trail toward Loch Leven, winding through a lumpy granite landscape dotted with Jeffrey pines. I'm heading for the spot where the granite forced itself into and through the terrane rock as it ballooned upward as magma 120 million years ago. My aim is to put my hands on the western wall of the great Sierra Nevada Batholith. To understand what's coming, I need to take you on an aside into the story of granite.

·   ·   ·

Granite is so emblematic of the high Sierra Nevada that some people assume that the range is made entirely of it. They are only a little wrong. Granite[1] lies exposed over about 15,000 square miles of the Sierra Nevada, and seismic surveys show that it may reach down close to 20 miles in places. Most of it is Cretaceous in age, with the peak of formation taking place between about 125 million and 80 million years ago.

Granite is handsome rock, crystalline and sparkling, light in color, like salt mixed with a bit of pepper and cinnamon. A close look reveals a gleaming collage of interlocking feldspar, quartz, mica, and amphibole minerals. In the eighteenth century, the renowned German mineralogist Abraham Werner argued that most rocks—granite included—precipitate out of seawater. The Scottish geologist James Hutton (appendix I), was swimming against Werner's considerable influence, and the prevailing wisdom of the day, when in 1795 he argued that granite forms from magma. Hutton got it right, but Werner wasn't all wrong. Seawater, it turns out, is essential for the formation of granite—although for reasons that neither Werner nor Hutton (both of whom lived long before the discovery of plate tectonics) could ever have imagined.

Granite is born from magma that wells up where subduction consumes the ocean floor. Where oceanic plates subduct, they take quadrillions of tons of old seawater down with them into the mantle. (The seawater is trapped in the hydrated minerals of serpentinite below the oceanic crust, as well as in the mud of the seabed and in fractures in the crust.) Once a subducting plate reaches a depth of about eighty miles, the intensifying heat drives this trapped seawater out of the plate into the surrounding mantle. Water has a curious effect on hot mantle rock—it lowers its melting point. This might seem odd, but experiments show that if you heat up mantle rock close to its melting point and then add some water, the rock melts. This happens because water molecules, backed by the intense heat and pressure of mantle depths, interfere with the chemical bonds between atoms in rock, causing the rock to melt at lower temperatures. (The effect is not unlike scattering salt on an icy road; the salt interferes with the chemical bonds that form ice crystals, thus causing the ice to melt.) Catalyzed by water, the hot mantle just above the subducting oceanic plate begins to melt. This magma is the starting point of granite.

This new-formed magma, being less dense than the mantle around it, begins to float upward. It ascends in colossal molten blobs that may—at least in our idealized conception—look a bit like rising hot air balloons. The blobs wedge their way upward along fractures, melt their way through the overlying rock, or excavate their way by stoping—a process in which magma moves upward by detaching and engulfing pieces of the overlying rock. (Stoping—rhymes with groping—is a mining term that refers to extending a mine shaft upward by excavating the overlying ore rock.) On its upward trip, the magma undergoes a profound chemical change. It preferentially picks up lighter chemical elements (oxygen, silicon, sodium, and aluminum, in particular), because these melt most readily from the surrounding rock. Meanwhile, some of the heavier elements in the magma (iron and magnesium, in particular) form crystals that drop away through the molten mass like sunken ships. By the time the magma has ascended to within a few miles of the Earth's surface, it has become chemically enriched in light elements and depleted of heavy ones. The key is this: upon cooling, this chemically evolved magma will form rock very different from the mantle out of which it was born. It will form—among other things—granite.[2] The granite of the Sierra Nevada formed this way, from the subduction of the Farallon Plate (figure 4.1).

Some of the magma rising from the Farallon Plate forced its way up to the Earth's surface. There it exploded across the Cretaceous world, incinerating unlucky dino-

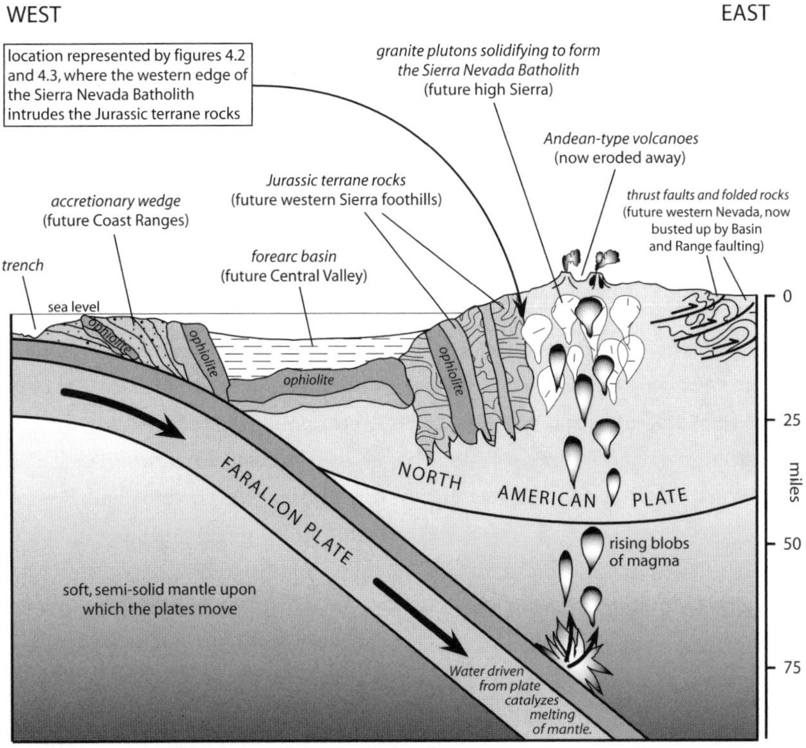

WEST   EAST

location represented by figures 4.2
and 4.3, where the western edge of
the Sierra Nevada Batholith
intrudes the Jurassic terrane rocks

granite plutons solidifying to form
the Sierra Nevada Batholith
(future high Sierra)

Andean-type volcanoes
(now eroded away)

accretionary wedge
(future Coast Ranges)

Jurassic terrane rocks
(future western Sierra foothills)

thrust faults and folded rocks
(future western Nevada, now
busted up by Basin
and Range faulting)

trench

forearc basin
(future Central Valley)

sea level

ophiolite

ophiolite

ophiolite

ophiolite

0

25

miles

FARALLON PLATE

NORTH   AMERICAN   PLATE

50

rising blobs
of magma

soft, semi-solid mantle upon
which the plates move

75

Water driven
from plate
catalyzes
melting
of mantle.

FIGURE 4.1

The origin of the Sierra Nevada Batholith. Subduction of the Farallon Plate during Cretaceous time generated magma that rose in great blobs to form granite plutons, mostly between 125 million and 80 million years ago. Hundreds of plutons glommed together to create the Sierra Nevada Batholith. Some of the rising magma erupted as volcanoes, which erosion subsequently erased as it cut down to the granite beneath. Today's Andes represent a modern version of the plate tectonic setting portrayed here. The Andean volcanic chain rises where the oceanic Nazca Plate subducts under South America at the Peru-Chile Trench. If you could cut several miles off the top of the Andes, you would see fresh granite like that in the Sierra Nevada, as well as glowing magma on its way to becoming yet more granite. Note the location represented by the outcrops shown in figures 4.2 and 4.3.

saurs while hurling forth clouds of sky-darkening ash and piling up into tall, Andean-style volcanic cones. These Cretaceous volcanoes once stood high over where the Sierra Nevada is now, but erosion has since taken them away. Most of the magma, however, didn't make it to the surface. It gurgled to a stop some five to ten miles underground, where it began to cool into granite. The atoms in the molten mass began to slow down and feel their mutual attraction. Atoms of opposite electrical charge began to join up in ionic bonds. Those with electrons to share began to join up in covalent bonds. By the trillions and quadrillions, they joined up, gathering into precise rows and columns like soldiers assembling for a grand review. The result: geometrically exact crystals of quartz, feldspar, mica, and hornblende—the mineral building blocks of granite.

·   ·   ·

Back now on the Loch Leven trail, I bash open a piece of granite with my hammer and turn the broken surface to the light. It flashes like a disco ball as sunbeams ricochet off the mineral cleavage planes. Only deep underground does magma cool slowly enough to make such large, geometrically precise crystals. Magma that erupts as lava solidifies so quickly that the atoms have time to form only small crystals. Run your fingers over a piece of lava rock; the sandpaper texture shows you how big the crystals were able to grow before the lava solidified—crystals rarely bigger than table salt. But a stately pace of cooling deep underground gives atoms time to gather together into large crystals. Peppercorn- to fingernail-sized crystals are typical in granite, although with a bit of searching you'll often find crystals as big as your thumb or even your fist.

Where blobs of molten granite stall deep underground, they solidify into roughly spherical bodies up to several miles across called plutons. Multiple plutons join together into batholiths: immense granite masses tens of miles wide and hundreds long. If a marshmallow represents a pluton, a batholith is like a bag of old marshmallows all fused together. The granite plutons of the Sierra Nevada (which number at least several hundred) comprise the Sierra Nevada Batholith. My hike along the Loch Leven trail is taking me across the eroded core of one of these plutons, known as the Rattlesnake Creek Pluton. It solidified about 120 million years ago at the very western edge of the batholith. The Rattlesnake Creek Pluton is, in effect, one of the outermost marshmallows in the bag. And that's why I'm here. Because when I locate the edge of the Rattlesnake Creek Pluton, I will be standing at the edge of the great batholith.

Everywhere along the trail, the granite is sprinkled with irregular pieces of dark terrane rock, some the size of golf balls, others as large as footballs. As granite-forming magmas punch their way up through the Earth, they collect chunks of the surrounding rock (known as country rock) and bring them along for the ride. As the magma solidifies into granite, these chunks of country rock get trapped within the granite like pieces of fruit in congealing Jell-O. Xenoliths—literally "foreign rocks"—is the term given to these dispossessed pieces of country rock. It's an odd term considering that the granite is really the foreigner: the invader that displaced the native rock of the region. As I continue along the trail, the xenoliths become more abundant—and *bigger*. Footballs become beach balls, beach balls become taxies, and taxies begin to crowd together. Within a few yards, I arrive at the edge of the Rattlesnake Creek Pluton—the place where the magma could no longer make headway against the country rock. I've arrived at the western wall of the Sierra Nevada Batholith (figure 4.2).

West from where I stand to the Sierra Nevada foothills, the bedrock is mostly Jurassic terrane rock. East to Donner Pass and a bit beyond, it is Cretaceous granite. The contact that divides these rocks—the wall of the batholith—represents part of an immense cliff

EAST
→
to Donner Pass

WEST
→
to Sacramento

terrane rock (remains of a ~200-million-year-old volcanic island arc)

granite of Rattlesnake Creek Pluton (~120 million years old)

FIGURE 4.2

Where the western edge of the Sierra Nevada Batholith (represented here by the Rattlesnake Creek Pluton) meets the Jurassic terrane rock (the country rock into which the granite-forming magma intruded). This boundary, or contact, extends north and south for many miles through the Sierra Nevada and probably reaches more than ten vertical miles down. The boundary is not as sharp as implied in this restricted view. Rather, it constitutes a zone up to several hundred feet wide where the granite and the terrane rock intermix. Located along the foot trail to Loch Leven near Big Bend in the Yuba River Valley.

that plummets more than ten miles below my feet. The wall also once reached several miles overhead, but erosion has since taken that part away. Spin the clock back 120 million years, and you see not granite here but molten rock, sloshing in viscous waves against the wall of an immense underground chamber. Xenoliths break free from the wall and gyrate slowly in the molten eddies. Fingers of liquid granite probe the wall, seeking to enlarge the chamber. Today, these granite fingers lie frozen in time, trapped in the act of prying fresh xenoliths from the country rock (figure 4.3).

I turn and head back down the trail. Granite gravel crunches underfoot. It is a wondrous thing to walk on granite—a rock that forms five or more miles underground.

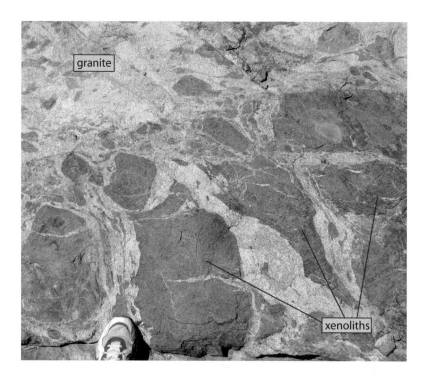

FIGURE 4.3

Fingers of granite caught in the act of kidnapping xenoliths at the edge of the Sierra Nevada Batholith. Notice how some of the xenoliths fit together like jigsaw puzzle pieces, separated only slightly by the intruding granite. The toe of my sneaker gives scale.

The only way that granite can see daylight is if erosion strips away the rock above it. Cut five to ten miles off the top of the Andes, and you will find granite. Much of the world's granite is still deep underground. But in areas where tectonic forces have pushed up mountain ranges, erosion can lay bare the granite within. Uplift and erosion—twin forces whose work is on spectacular display up ahead, along the granite ridgeline of the high Sierra at Donner Pass.

I accelerate back onto Interstate 80 and continue east, climbing higher. "Climb the mountains and get their good tidings," John Muir preached. He would have been appalled at the way many of us follow his directive today, by stomping the accelerator pedal. The highway, ascending the broad sweep of the Yuba River Valley, enters the heart of the Sierra Nevada Batholith. Cretaceous granite spreads in all directions, cloaked with pines that sink their roots not into soil but cracks in the granite. There isn't much soil in the high Sierra Nevada; it hasn't had time to reform since glacial days. As recently at 12,000 years ago, groaning rivers of ice oozed east and west off the Sierra crest, scraping the soil from the valleys and cutting deeply into the underlying granite. The ice scooped out great amphitheater-like bowls, called cirques, at the heads of the glacial valleys and left the crest of the range adorned with sharp ridges and toothy summits. Small ice streams merged into colossal ice rivers that excavated immense gorges like Kings Canyon, Kern Canyon, and Yosemite Valley. Twelve thousand years ago is not much in geologic time. Compared to the age of the Earth, that's as recent as three hours ago is to the U.S. Civil War. The scratches of recent ice score the granite throughout the high Sierra (figure 4.4).

I exit the interstate and follow old Highway 40 up to Donner Pass, 7,090 feet above the sea. Here, finishing our traverse of Muir's range, we'll look at how ice sculpted its crest and, in so doing, shaped California's history. The pass forms a quarter-mile-wide saddle-shaped divide between high peaks to the north and south. Donner Pass sits on the great dividing ridge of the Sierra Nevada, where streams split for either the Pacific Ocean or the Great Basin. During the ice ages, the dividing ridge split glaciers. Flowing east and west off the ridge, the ice carved back-to-back cirques on both sides of Donner Pass and left its signature scratches all over the granite.

Climbing an ice-carved knoll on the west side of Donner Pass, I take in the vast sweep of the west-facing cirque. It forms a half-bowl more than a mile across and open to the west. Ski lifts radiate from the bowl's center up toward the lip. Westerly winds funnel up the cirque and sweep over the pass, pruning the scattered pines into permanent east-streaming weathervanes. Turning and walking east across the pass, I soon reach the rim of the east-facing cirque, which drops 1,160 feet to Donner Lake, a three-mile-long lozenge of cobalt blue. The ice, as it descended from the pass, plucked house-sized chunks of granite off the mountainside, leaving the route up to the pass barricaded by high, step-like ledges. The ice carried those chunks several miles to what is now the east end of Donner Lake, before dropping them to form a moraine—a ridge of rock debris that accumulates at the melting edge of a glacier. The moraine forms a natural dam that holds back Donner Lake.

It was on that moraine that most of the Donner party spent the winter of 1846–1847, trapped by snows up to twenty feet deep. Marooned for more than five months, the pioneers ate up all of their cattle and dogs. They boiled ox bones for broth and then reboiled the same bones again, and again. They boiled ox hides to make glutinous soups. Then they cut the hides into strips and ate those. Then, as some died, their bodies became calories for those still living. Despite heroic rescue efforts, thirty-six members of the party died of starvation or cold, and the last of the survivors did not reach safety until the end of April 1847. The evidence of the ordeal, still fresh, greeted many forty-niners who, coming from the east, passed through the remains of the starvation camps on their way up to Donner Pass. Wakeman Bryarly, wandering past the Donner's abandoned camps on August 21, 1849, claimed to have "found many human bones. The skulls had been sawed open for the purpose, no doubt, of getting out the brains, & the bones had all been sawed open & broken to obtain the last particle of nutriment." Likewise, forty-niner Charles Long reported "a great many human bones" lying about the camps. A. J. McCall, passing through on September 7, 1849, marveled at "stumps from

FIGURE 4.4

The marks of glaciers score the granite throughout the high Sierra Nevada. The crescentic gouges in the top image formed where ice-dragged boulders skipped across the rock, digging crescent-shaped divots whose convex side points in the direction that the ice flowed. Note sleepy spouse for scale. The striations in the bottom image formed where ice-dragged pebbles scratched and polished the underlying granite. Here as well, the flowing ice plucked a chunk of granite from the downstream side of a bulge in the rock, leaving a scar called a pluck mark.

ten to twenty feet high that were haggled off at the snow line," testifying to the cruel depths of snow that had sealed the Donner's fate. Few of these passersby lingered for long, though. They had their own ordeal to face: the fearsome ascent up to Donner Pass.

The ice-carved granite ridge of the Sierra Nevada determined the course of two major events in nineteenth-century U.S. history: the overland emigration to California and the laying of the transcontinental railroad. Both events converge at Donner Pass. In 1844, an emigrant party lead by Elijah Stephens took the first wagons west over what would later be called Donner Pass. By showing that wagons could conquer the Sierra Nevada, the Stephens party opened the emigration floodgates to California. But even though Donner Pass is one of the lowest Sierra Nevada passes, it was one of the hardest ways to cross by wagon because of the glacier-plucked granite ledges that block the approach from the east. "Standing at the bottom and looking upwards at the perpendicular, and in some cases, impending granite cliffs," 1846 emigrant Edwin Bryant wrote below Donner Pass, "the observer, without any further knowledge on the subject, would doubt if man or beast ever made a good passage over them."

Often, the only way to get up and over the ledges was to take the wagons apart and haul them up piece-by-piece (figure 4.5). Another strategy was to cut long poles from sapling trees and lay these up against the ledges, like laying long boards against a flight of stairs. The pioneers would then slide the wagons up the poles. It was brutal work, and a powerful incentive to find other options. During the 1846 emigration season, scouts reconnoitered a new pass, called Roller Pass, about one mile south of Donner Pass. Although nearly 800 feet higher than Donner Pass, the approach to Roller Pass has no high ledges. However, the final 400 feet form a 30-degree slope of jagged boulders—"as steep as the roof of a house," forty-niner Joseph Hackney observed. Oxen could not pull wagons directly up such a grade. Instead, the animals were unhitched and driven to the top, where they could heave along the flatter ground at the crest of the pass. The emigrants then yoked the oxen to hundreds of feet of chain and passed the chain over log rollers down to the waiting wagons (figure 4.6). Then, under shouts and cracking whips, as many as twelve yoke of oxen hove to on each wagon.[3] The taut chain gnashed into the log rollers, and one by one the wagons creaked upward to gain the crest of the mountain.

Between 1841 and 1869, more than a quarter-million emigrants arrived from the east to scrabble across the ridgeline of the Sierra Nevada into California. The peak of the emigration was from 1849 to 1853—the first five years of the gold rush. By then, arriving pioneers had largely abandoned the daunting ascents to Donner Pass and Roller Pass in favor of other routes, particularly Carson Pass about sixty miles to the south. A map of these old pioneer trails, traced from the Great Basin toward California, looks like the unraveling end of a rope, with the various strands splaying across the Sierra Nevada at strategic points.

Wagon train emigration to California came to a virtual halt in 1869, when the transcontinental railroad opened for business. Donner Pass made that possible. From 1866

FIGURE 4.5

Glaciers and granite combined to create brutal toil for nineteenth-century emigrants crossing the Sierra Nevada over Donner Pass. This painting, by Harold von Schmidt, illustrates the strategy for ascending the glacier-plucked ledges east of Donner Pass. Emigrant Benjamin Bonney in 1846 described the scene: "We came to a rim rock ledge where there was no chance to drive up, so the wagons were taken to pieces and hoisted to the top of the rim rocks with ropes.The wagons were put together again, reloaded and the oxen which had been led through a narrow crevice in the rim rock were hitched up and went on."

to 1867, Chinese laborers for the Central Pacific Railroad punched a one-third-mile-long tunnel through the granite below Donner Pass. Using only hand tools, black powder, and nitroglycerin, the Chinese dug the summit tunnel and blasted out notches in the mountainside to make the railroad bed. They bridged a section of sheer cliff just east of Donner Pass with a massive wall of cut granite blocks. Today's railroad passes through a newer tunnel, but the original tunnel and cut-block wall still stand at Donner Pass, testifying to the tenacity and skill of the Chinese. On May 10, 1869, after laying a combined 1,756 miles of track over six years, the Central Pacific line (heading east from Sacramento) and the Union Pacific line (heading west from Omaha) met in a sagebrush wilderness in

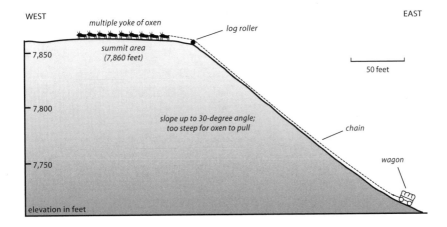

FIGURE 4.6
This figure shows—with no vertical exaggeration—the situation at Roller Pass. The ascent, up the headwall of a glacier-cut cirque, demanded the use of log rollers and multiple yoke of oxen on each wagon.

Utah called Promontory Summit. Travel time from the Missouri River to California suddenly dropped from one hundred days by ox-drawn wagon to *four* days by train—a difference so profound that it would not be matched, relatively speaking, until the age of jet travel. The cost dropped too. In 1870, an "Emigrant Class" ticket from Omaha to Sacramento sold for $60—roughly the cost of two good yoke of oxen during the gold rush.

NOTES

1. Throughout this chapter, I use the term "granite" in a broader sense than some specialists prefer. Much of the "granite" in the Sierra Nevada is technically granodiorite, tonalite, or quartz monzonite. The distinctions between these closely related members of the granite family are not important for our purposes.

2. This process, known technically as magmatic differentiation, is of supreme importance not just for the formation of granite but for other igneous (meaning magma-derived) rocks of the continental crust. Although the mantle is the ultimate source of most magma, a magma that becomes sufficiently differentiated from its mantle source will produce rock quite different from the mantle. Where such a magma erupts, it may make andesite and rhyolite lava flows and volcanic tuffs (compacted deposits of volcanic ash). Where it solidifies underground, it may make diorite, granodiorite, granite, and related rocks. These rocks comprise the main igneous rocks of the continental crust.

3. One yoke of oxen is two animals harnessed side-by-side with a yoke: a crossbar of carved wood fastened to the oxen's necks with oxbows.

5

WHERE IS THE EDGE OF THE
NORTH AMERICAN PLATE?

Reykjavík, the capital of Iceland, sits on the edge of Faxaflói Bay, a huge cove that faces southwest to snare some warmth from the northern tendrils of the Gulf Stream. The surrounding tundra landscape is stark, dominated by dark lava beds and volcanic cones. If lush vegetation is what you seek, cross Iceland off your list. A sad collection of mosses and shriveled shrubs clings to cracks in the lava beds. Iceland is built entirely of lava, most of it less than three million years old. Magma hovers close underground nearly everywhere, making Iceland a geothermal tourist's delight. Within an hour of landing in Reykjavík, my wife Susan and I plunged into the steaming Blue Lagoon, where we soaked away the cramps of the long flight from San Francisco. Over the next few days, our explorations would take us to dozens of steam vents, hot pools, and geysers, including the one at Geyser, from which all others take their name. Our fondest Icelandic memories involve geothermal bathing. Icelandic showers have heads a foot across, with holes big enough to stick in a large nail. To get water, you don't twist a little knob; you haul on a big lever, and brace yourself for a deluge. It's all guilt-free, for the water is warmed not by climate-changing fossil fuels but by hot rock underground. Electricity, home heating, and hot water: Icelanders get them all practically free from geothermal heat. It's one of the benefits of living on the edge—the plate edge, that is.

Icelanders look to Europe for their ancestry and cultural identity. But geologically, their island is evenly divided between Europe and North America. Iceland sits on the Mid-Atlantic Ridge—the spreading, splitting boundary between the Eurasian and North American tectonic plates (see the frontispiece tectonic plates map). In most places, the

THINGVELLIR RIFT VALLEY
(~10 to 50 miles wide; Eurasian
Plate begins on the other side)

NORTH AMERICAN PLATE
(~5500 miles to the San Andreas
fault and the Pacific Plate)

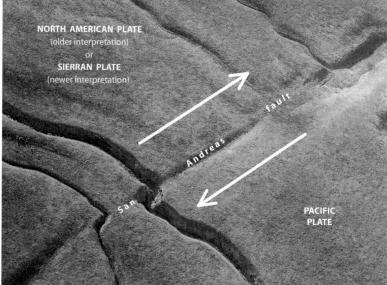

NORTH AMERICAN PLATE
(older interpretation)
or
SIERRAN PLATE
(newer interpretation)

San Andreas fault

PACIFIC
PLATE

FIGURE 5.1

The opposite edges of the North American plate—at least by traditional definition. The top photograph looks south along the west edge of the Thingvellir Rift Valley, Iceland's great rift zone where the Mid-Atlantic Ridge rises above the sea. If the North American Plate has a definitive eastern edge, I'm standing on it. The bottom photograph is an aerial view of the San Andreas fault in California's Carrizo Plain. The linear crease and offset stream channels mark the fault and show how it moves. Although tradition asserts that the San Andreas fault represents the North American Plate's western edge, this chapter argues that the plate—in the southwestern United States at least—has *no* true edge.

Mid-Atlantic Ridge lies a mile or more below sea level, running north–south through the middle of the Atlantic Ocean like a great zipper—which, in some sense, it is: the zipper that opened the Atlantic. The ridge rises out of the sea at Iceland because there, apparently, it runs across a mantle plume: an intense point-source of magma rising from the Earth's deep interior. Thanks to the overlap of spreading ridge and mantle plume, lava has heaped up so thickly here that the Mid-Atlantic Ridge rises above the sea to make Iceland. The center of the ridge slashes directly through Iceland, forming a rift valley ten to fifty miles across (figure 5.1). Iceland splits and grows from this central rift valley as lava oozes forth periodically. Reykjavík, west of the rift valley on the North American Plate, is creeping steadily west away from Hvolsvöllur, east of the rift valley on the Eurasian Plate. Each year, the drive between the two cities increases by about three-quarters of an inch—as cracks in the connecting highway and GPS measurements attest.

Californians, like Icelanders, also live along a plate edge. Unlike Icelanders, Californians reap few benefits from the association—other than the thrill of knowing that death from an earthquake could come any day, which naturally makes each latté and yoga lesson more meaningful. Iceland and California lie at opposite edges of the North American Plate. The Mid-Atlantic Ridge forms the plate's eastern edge. In California, the San Andreas fault forms its western edge (figure 5.1). At least, that's what the textbooks all say. Look up the San Andreas fault in any geology book, and you'll read that it forms the boundary between the North American and Pacific plates. The problem with this well-entrenched fact is that, although it's not quite wrong, it's not quite right either. Moreover, it obscures a more interesting story about the dynamic geology of the American West. In truth, as we have learned more about the West, we have actually become *less* certain about the western edge of the North American Plate. To understand what I mean, we need to learn a few things about earthquakes, and then go visit the San Andreas fault. I hope to convince you that the San Andreas fault is just one of several big players in the plate-moving game. The implications include some novel ideas, such as the notion that the Sierra Nevada and Central Valley together form their *own* tectonic plate (only recently recognized), and that ocean waves may one day break in the deserts of Nevada.

.   .   .

Earthquakes are the Earth's way of releasing the stresses that build where tectonic plates grind past one another. Ask any geologist what a "plate" is, and you'll hear an explanation that includes earthquakes. Plates, they'll say, are horizontally moving blocks of the Earth's lithosphere[1] whose edges chafe against neighboring plates to produce earthquakes. That's why a world map of earthquakes translates directly into a map of the Earth's tectonic plates (figure 5.2).

Most earthquakes occur in the ocean basins, along mid-ocean ridges and ocean trenches—the Earth's major plate boundaries. But not all plate boundaries lie in the

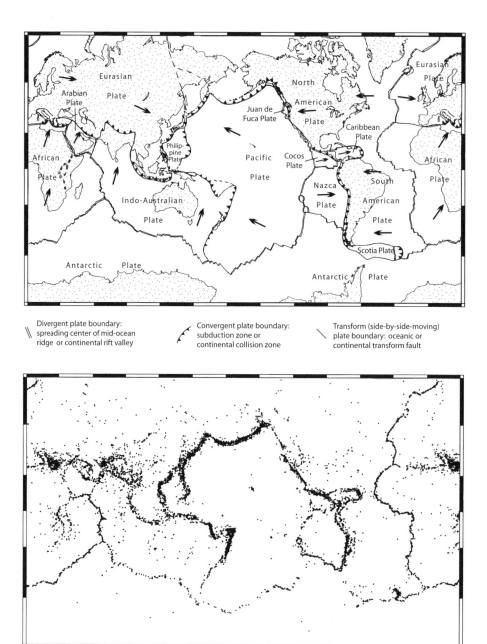

Divergent plate boundary: spreading center of mid-ocean ridge or continental rift valley

Convergent plate boundary: subduction zone or continental collision zone

Transform (side-by-side-moving) plate boundary: oceanic or continental transform fault

FIGURE 5.2

The top map shows the Earth's major tectonic plates, with the arrows indicating the direction of plate movement. The bottom map shows global earthquakes recorded over a two-decade period. You can see that most earthquakes occur along plate edges. (The data set is limited to quakes with depths of 40 miles or less and magnitudes of 4.5 or more.)

ocean basins. Some, like the San Andreas fault, slash across portions of the continents. The San Andreas marks the place (at least in part) where the North American and Pacific plates slide side-by-side past one another. But the plates don't slide smoothly. Friction locks them together. As the plates try to slide, stress builds. The rocks on each side of the fault bend to take up the strain. Eventually, the rocks reach their breaking point. The fault snaps, and the rocks on each side leap violently to a new position. The energy released surges through the Earth in pulsing spasms called earthquake waves.

If the San Andreas fault were to shift a few inches right now, the resulting earthquake would be tiny. If you were near the epicenter, the waves would rattle cups and glasses in your cupboard and cause a momentary lapse in attention to *The Simpsons*. "Did you feel that?" If the fault were to shift several feet, a moderately large quake would result. Houses would sway, plaster walls might crack, standing bookshelves might pitch over, and items would fall off the mantelpiece. Neighbors would check on one another. "Is everyone OK?" If the fault were to leap ten feet or more, it would likely spell catastrophe even miles from the epicenter. Houses would slide off foundations, cars would plunge off collapsing freeway overpasses, bodies would lie crushed under rubble.

On January 9, 1857, the San Andreas fault leapt *thirty feet* in a virtual instant as it ruptured along a 225-mile seam from Cajon Pass to Parkfield (figure 5.3). People outdoors fell over and clutched the ground in terror as the Earth heaved in sickening waves. It was "similar to sea sickness," one witness wrote. "Everything (houses, trees, cattle and people) had the appearance of being drunk." The injured and homeless sought shelter at Fort Tejon, a U.S. Army encampment north of Los Angeles (figure 5.3), only to find the fort in ruins and the soldiers shivering in makeshift tents. Newspapers and diarists reported the shaking as far away as San Diego, Sacramento, Las Vegas, and Yuma. Although it was the strongest quake ever witnessed (so far) in California, the 1857 Fort Tejon quake wasn't particularly deadly, thanks to the state's sparse population. That would not be the case on April 18, 1906. On that day, the San Andreas jumped up to twenty feet along a 300-mile stretch centered on San Francisco. Within minutes, most of San Francisco lay in ruins, and fires over the next days multiplied the devastation.

Since the 1906 San Francisco earthquake, the San Andreas fault has unleashed only modest tremors. (The Loma Prieta quake of October 17, 1989—sometimes called the World Series quake because it happened during the third game at Candlestick Park—occurred not on the San Andreas but on a previously unknown fault nearby.) Yet the Pacific Plate is still sliding northwest relative to the North American Plate by about 2.0 inches per year—a number first determined from geologic evidence and recently affirmed by GPS measurements.[2] How, and where, is this colossal tectonic stress being absorbed? Is the San Andreas fault, like a stretching rubber band, getting ready to snap somewhere and unleash another calamitous blast? Perhaps. But the San Andreas fault isn't the only player in the game. Although the San Andreas has stayed rather quiet for more than a century now, other faults from California to Nevada have been making plenty of noise. As I'll show you, when we go and measure how much the San Andreas

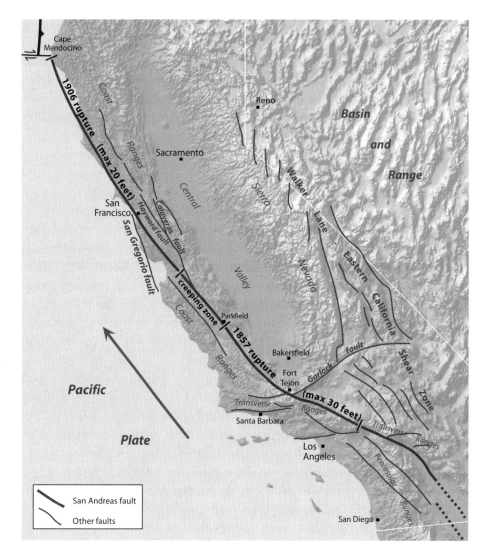

FIGURE 5.3
California's big faults, showing the segments of the San Andreas fault that ruptured during the 1857 Fort Tejon and 1906 San Francisco earthquakes.

fault is actually moving, we find that we need those other faults to balance the books. Only by including the other faults can we come up the full 2.0 inches per year that the North American and Pacific plates are moving past one another. This has significant implications, both for the seismic hazards of today and the geography of the western United States in the geologic tomorrow.

The San Andreas fault is rarely hard to find. Erosion excavates the pulverized rock along the fault zone, so that a narrow valley scribes the fault nearly everywhere. From the air, the mark is as clear as the centerline on a highway. Flying between Los Angeles and San Francisco, you could find your way by the fault. About one-third of the way from L.A. to San Francisco, a look out the airplane window shows the fault slashing a straight line across a desert basin called the Carrizo Plain. In its wild remoteness, the Carrizo Plain epitomizes a geographic truth about California. This state—caricatured by clichés involving surfing, traffic, and urban sprawl—is mostly empty wilderness. For every square mile of the urban Los Angeles and San Francisco Bay areas combined, California boasts more than twenty square miles of wilderness deserts, mountains, and forests. For much of the year, the Carrizo Plain is a hot, windswept desert that sees few visitors. But in winter, rains soften the plain and the nearby mountains with wildflowers and new-sprung grass. If enough rain falls on the Temblor Range, a large arroyo called Wallace Creek flows west from the range onto the Carrizo Plain. No one but a geologist—in this case Robert Wallace, pioneering researcher of the San Andreas fault—would be honored to have his name affixed to a gravelly, weed-choked arroyo in the middle of nowhere. But Wallace Creek is not just any gravelly, weed-choked arroyo, for it crosses the San Andreas fault. Sidling movements along the fault have offset the creek's channel so that it now forms the slash-of-Zorro bend that you can see in figure 5.4.

Whenever I visit the San Andreas fault at Wallace Creek, I do what everybody does: straddle the well-exposed trace of the fault and delight in having one foot on the North American Plate and the other on the Pacific Plate. One night I slept on the fault, laying my bedroll in the crease that it forms on the landscape near that pair of offset channels that you can see in figure 5.1. I looked up at the stars with one outstretched arm linked to Reykjavík on the North American Plate and the other linked to Honolulu on the Pacific Plate. Whether it was the bourbon or the enchantment of the moment, I don't know, but I could almost feel the two plates straining under my shoulders, trying to go their separate ways. Alas, my reverie could not last. I knew that the San Andreas fault is not, in fact, the boundary between the North American and Pacific plates—not exclusively anyway. To see the evidence, I knew I needed to look no further than Wallace Creek.

Where Wallace Creek meets the fault, it angles northwest and follows the fault for 420 feet before turning away. If you trace the fault northwest from this point, you come to an older, now-abandoned channel of Wallace Creek (figure 5.4). Both channels, when they first formed, probably went straight across the fault. If we can figure out when each channel formed, we can use their sideways displacements to calculate the long-term rate of movement on the San Andreas fault. Cal Tech geologist Kerry Sieh and colleagues have done this by trenching sediments in the channels and age dating (using radioactive carbon) plant matter trapped in those sediments. I've summed up Sieh's results in the bottom diagrams of figure 5.4. The older, now-abandoned channel of Wallace Creek appears to have formed about 11,000 to 10,000 years ago (5.4a). Earthquakes along the fault periodically displaced this channel, forcing the water to flow along an

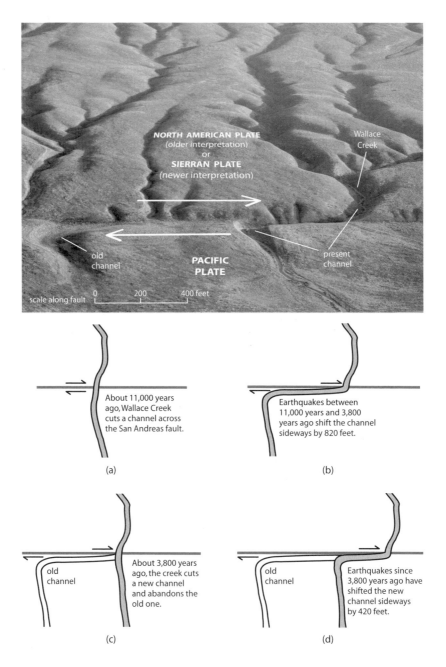

(a) About 11,000 years ago, Wallace Creek cuts a channel across the San Andreas fault.

(b) Earthquakes between 11,000 years and 3,800 years ago shift the channel sideways by 820 feet.

(c) About 3,800 years ago, the creek cuts a new channel and abandons the old one.

(d) Earthquakes since 3,800 years ago have shifted the new channel sideways by 420 feet.

FIGURE 5.4

The photograph (top) looks northeast across the San Andreas fault at Wallace Creek in the Carrizo Plain. When Wallace Creek flows, the water runs toward you out of the Temblor Range. Notice the dogleg offset of both the present channel and an old, now-abandoned channel of Wallace Creek. Illustrations (a) through (d) show how today's scene came to be (bottom).

ever-lengthening dogleg. By 3,800 years ago, the cumulative work of dozens of earthquakes had offset the channel by 820 feet (5.4b). Wallace Creek then punched a new channel straight across the fault, probably during a flood, and abandoned its old channel (5.4c). Every earthquake since this new channel formed 3,800 years ago has added to the displacement of this channel, which is now offset by 420 feet (5.4d).

If you get out your calculator and divide the offsets of the two channels by their ages, you will obtain an average rate of movement of 1.35 inches per year for the San Andreas fault over the last 11,000 years. The fault, of course, doesn't actually move 1.35 inches each year—that's just the long-term average. Rather, it leaps in spasmodic jerks several inches or several feet at a time, and then sits still for many years. (By analogy, think about the average speed of your car over its lifetime, taking into account both the short periods when it's moving and the long periods when it sits in the driveway.) As I mentioned, the total motion between the Pacific and North American plates is 2.0 inches per year. Conclusion: the data from Wallace Creek suggest that the San Andreas fault, at least in recent millennia, has absorbed only about two-thirds of the motion between the two great plates. This must mean that other faults, elsewhere, are contributing to the movements. But just in case you think Wallace Creek is a fluke, let me support the results from there with examples from other sections of the San Andreas fault. I'll give you three.

About two million years ago, boulder-laden floods poured west off the flanks of Loma Prieta Mountain (about ten miles south of present-day San Jose), strewing bouldery deposits across the San Andreas fault. Movements along the fault have since split these boulder deposits in two and shipped them away from one another. To match up the boulder beds today, you have to go twenty-three miles northwest of Loma Prieta Mountain to Los Trancos Preserve in the Santa Cruz Mountains. Twenty-three miles divided by two million years yields an average rate of movement of 0.73 inches per year for the fault—even less than the 1.35 inches per year documented at Wallace Creek.

Figure 5.5 shows a second example. The Neenach volcanic field, a few miles east of Tejon Pass in the Transverse Ranges, consists of volcanic rocks that match like twins those of the Pinnacles volcanic field, on the west side of the San Andreas fault in central California east of Monterey. You can see on figure 5.5 that the San Andreas has split this once-continuous stack of volcanic rocks and shipped them to points that are now 195 miles apart! Radiometric dating (see appendix I for an explanation) shows that the volcanics erupted about 23.5 million years ago. However, the complexities of the San Andreas' evolution in the Transverse Ranges may mean that the fault didn't begin to split the volcanic rocks until perhaps 16 million years ago. Depending on which age you use, distance divided by time gives an offset rate of 0.53 to 0.77 inches per year—again, less than the 1.35 inches per year that we see at Wallace Creek.

A third and final example takes us to southern California. In the Chocolate Mountains east of the San Andreas fault near the Salton Sea lies a distinct suite of volcanic rocks that streams eroded and washed west as alluvial fans about twelve to thirteen million years ago. To find the portions of these alluvial fans that spilled onto the west side of the San

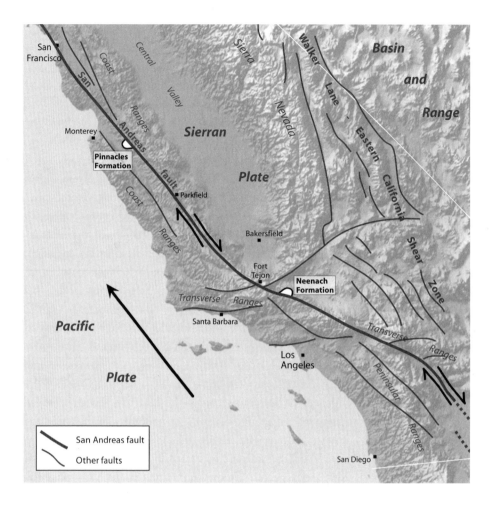

FIGURE 5.5

The San Andreas fault has split and separated many rock units along its length, but perhaps none more dramatically than the Pinnacles and Neenach volcanic formations. This once-continuous stack of volcanic tuffs and lava flows was torn in two by the fault, so that the two sections now lie 195 miles apart on opposite sides of the fault.

Andreas fault, you have to go 135 miles northwest to the Transverse Ranges north of Los Angeles. Distance divided by time in this instance yields an offset rate of 0.69 inches per year—once again, less than the 1.35 inches per year that we see at Wallace Creek.

More examples won't change the main conclusion that we can draw from the ones that I've just given. Everywhere we look, the San Andreas fault comes up well short of the 2.0 inches per year needed to account for the movement between the North American and Pacific plates. At best, the fault accounts for two-thirds of that movement, and possibly quite a bit less.

Hold a deck of cards vertically between your palms. The palm of your left hand represents the Pacific Plate, and the palm of your right the North American Plate. Slide your left hand away from you, while holding your right hand in place. The cards slide past one another to accommodate the motion. The pair of cards that slides the most represents the rocks on either side of the San Andreas fault. But many pairs of cards sliding past one another contribute to the overall motion. Likewise, slippage along the San Andreas fault is, in most places, accommodated not by the master fault alone but by many roughly parallel faults spread out across a ten- to eighty-mile-wide zone called the San Andreas fault system. In the Bay Area, for instance, the fault system includes (from west to east across a forty-mile-wide swath) the San Gregorio, San Andreas, Hayward, Rogers Creek, Calaveras, Concord-Green Valley, and Greenville faults, all of which are active or have been active in the recent geologic past. If we add up *all* of the motion on *all* the faults in the whole San Andreas fault system, we come up with about 1.5 inches per year of total motion. So we still haven't balanced the books; there's still a half-inch per year of motion to account for to reach that 2.0 inches per year of total motion between the Pacific and North American plates. To find that final half-inch, we need to leave the San Andreas fault system, and head east toward Nevada.

The faults of the San Andreas system crackle with near-constant tremors; seismographs commonly detect more than a dozen earthquakes per day, most of them too tiny for you or me to feel.[3] But as we leave the San Andreas system and head east across the Central Valley and the Sierra Nevada, the seismicity fades as if you had closed the door to a noisy room. If the North American Plate actually ended (or began) at the San Andreas fault, then the seismic silence should last all the way to Iceland. But it doesn't. Instead, the seismicity begins to crackle anew where we drop off the Sierra Nevada's eastern ridge-line into the Basin and Range Province. Here we come to the Walker Lane–Eastern California Shear Zone, another belt of active earthquake faults. (I'll call it the Walker Lane system for short; you can see it labeled on figure 5.3.) Like the San Andreas system, the faults of the Walker Lane system, as they snap and jostle, let the rocks on the west side slide northwest relative to the rocks on the east side. In other words, the faults of the Walker Lane system let the Sierra Nevada and Central Valley slide northwest relative to Nevada and the rest of North America.

When an axe splits a log, it is no longer one chunk of wood. When an active fault system splits a plate, it is no longer one plate. Every earthquake along the Walker Lane system reinforces one conclusion: *North America contains another plate.* I've labeled it on figures 5.5 and 5.6 as the Sierran Plate. It qualifies as a separate plate because it moves on its own, not entirely with either the Pacific Plate or the North American Plate. The Sierran Plate's existence is both revealed and made possible by the active fault systems that flank it: the San Andreas system on its west side and the Walker Lane system on its east side. When we add up the motion on all the faults in *both* the San Andreas system and the Walker Lane system, we come up with nearly 90 percent (close to 1.8 inches per year) of North American Plate–Pacific Plate motion. So we still haven't balanced the books; we still

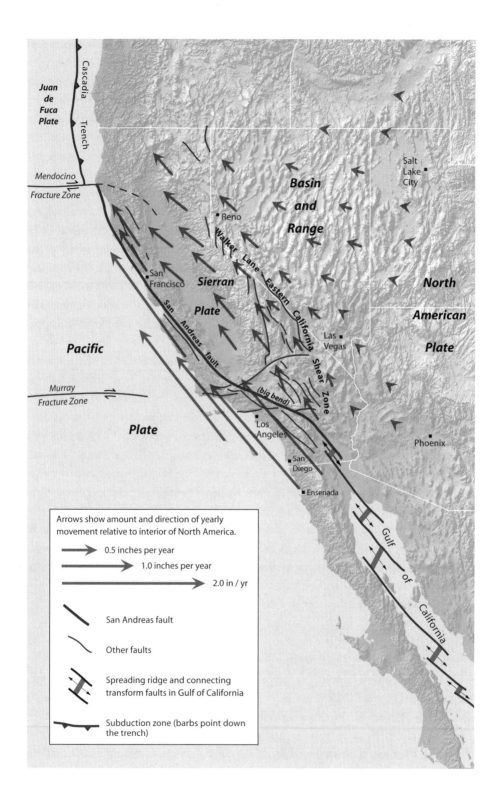

Juan
de
Fuca
Plate

Cascadia Trench

Mendocino
Fracture Zone

Pacific

Murray
Fracture Zone

Plate

San Francisco

Sierran

Plate

San Andreas fault

Reno

Walker Lane - Eastern California Shear Zone

(big bend)

Los Angeles

San Diego

Ensenada

Basin

and

Range

Salt Lake City

Las Vegas

North

American

Plate

Phoenix

Gulf of California

Arrows show amount and direction of yearly
movement relative to interior of North America.

0.5 inches per year

1.0 inches per year

2.0 in / yr

San Andreas fault

Other faults

Spreading ridge and connecting
transform faults in Gulf of California

Subduction zone (barbs point down
the trench)

need to find 10 percent more slip-sliding to reach that 2.0 inches per year of total motion. To locate this last bit, we need to head still farther east, into Nevada.

In recent years, global-positioning technology has allowed us to measure locations on the Earth's surface with astonishing precision. We can now measure fraction-of-an-inch movements between points that lie hundreds or thousands of miles apart, and thus track plate movements on a yearly basis.[4] Figure 5.6 shows some results. The arrows in the figure indicate how fast various places in the western United States are moving relative to the interior of North America. The lengths of the arrows reflect rates of movement; longer arrows mean faster movement. Notice on figure 5.6 that there are effectively three velocity clusters (groups of arrows with roughly equal directions and lengths). The first cluster lies on the Pacific Plate, showing that it moves northwest at about 2.0 inches per year relative to the interior of North America. The second cluster lies on the Sierran Plate, showing its northwesterly movement at a little over one-half inch per year. The third velocity cluster spreads across the Basin and Range Province. Points here are moving northwest at less than a half-inch per year, with the highest velocities near the Walker Lane fault system and decreasing eastward.

The thing that boggles my mind most about the data in figure 5.6 is how far east you have to go before you can stand on rock that *isn't* moving northwest relative to the rest of North America. The message is clear—the western 500 miles of the southwestern United States are slowly tearing away from the rest of the continent. Where is the edge of the North American Plate? You make the call. One thing is certain—it's not the San Andreas fault. The way I see it, the plate has *no* true edge in the southwestern United States. Like a vast ice sheet crumbling in warming seas, the continent is breaking up across a 500-mile-wide swath of active faults that stretches from the California coast to Salt Lake City. One manifestation of this great fragmentation is the Basin and Range Province. Here, the rhythmically spaced, north–south-trending mountain ranges and intervening valleys (basins) reflect as much as *250 miles* of east–west-directed continental stretching and fragmentation—a topic that we'll explore in chapters 6 and 7.

FIGURE 5.6 (OPPOSITE)

The North American Plate has no true edge in the southwestern United States because large portions of the plate are tearing away across a broad zone. The lengths of the arrows reflect the yearly rates of movement of particular points relative to the interior of North America, based on GPS satellite measurements. Regions west of the San Andreas fault, which include Baja California as well as the San Diego and Los Angeles areas, are all part of the Pacific Plate and are moving northwest at about 2.0 inches per year. The Sierran Plate, which lies between the San Andreas and Walker Lane fault systems, moves northwest a little over a half-inch per year. The Sierran Plate may be breaking free thanks, in part, to the "big bend" in the San Andreas fault, which lets the Pacific Plate push like a shoulder against the Sierran Plate's southern end. Compression along this big bend has pushed up the Transverse Ranges (labeled on figure 5.5), which are among the fastest-rising mountains in the West, popping upward at nearly a half-inch per year. The Basin and Range Province has already stretched by some 250 miles since its inception about twenty million years ago, and it continues to widen today by nearly a half-inch per year. As the western United States breaks apart, one or more ocean basins may eventually open up through the southwestern deserts, as shown in figure 5.7.

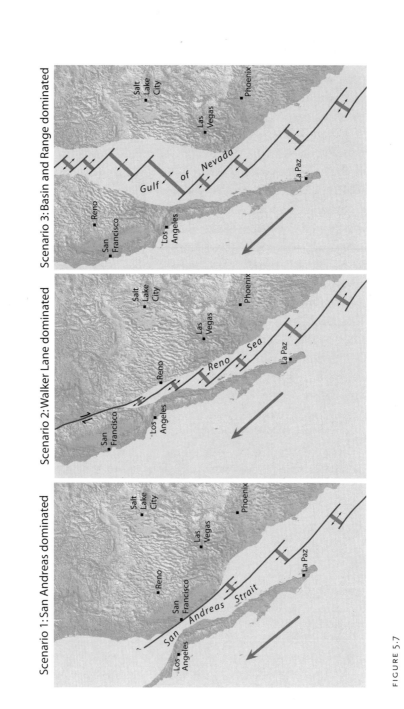

Scenario 1: San Andreas dominated

Scenario 2: Walker Lane dominated

Scenario 3: Basin and Range dominated

FIGURE 5.7

Possible geographies of the American West some fifteen million years in the future, assuming that present trends continue.

What are the implications of this ongoing fragmentation for the geography of the American West? In figure 5.7, I propose three not-too-fanciful scenarios. Each portrays a scene about fifteen million years in the future. Each assumes that the Pacific Plate will continue to move northwest at about 2.0 inches per year relative to the interior of North America.

In scenario 1 on figure 5.7, the San Andreas fault system takes over as the sole locus of motion. Baja California and coastal California shear away from the rest of the continent to form a long, skinny island.[5] A short ferry ride across the San Andreas Strait connects Los Angeles to the Bay Area.

In scenario 2, the San Andreas system sputters out, and the Walker Lane system takes over. All of California west of the Sierra Nevada, together with Baja California, shears away to the northwest. The Gulf of California extends like a wedge along the rift created by the Walker Lane to become the Reno Sea, which divides California from western Nevada. Residents of Nevada gambol and gamble along the shores of their new-formed ocean. The scene is reminiscent of how the Arabian Peninsula split from Africa to open the Red Sea some five million years ago.

In scenario 3, central Nevada splits open through the middle of the Basin and Range Province, where the highly stretched crust is already thin and weak. The widening Gulf of Nevada divides the continent from a large island composed of Washington, Oregon, California, Baja California, and western Nevada. The scene is akin to Madagascar's origin when it split from eastern Africa to open the Mozambique Channel.

Each scenario represents an end-member possibility; the actual result may well combine all three. But every likely projection points to one conclusion: continental fragmentation and eventual beachfront property in the deserts of the American West.

NOTES

1. The lithosphere is the Earth's outer shell of hard, rigid rock that forms the tectonic plates. It is about fifty to one hundred miles thick and includes the crust plus the uppermost part of the mantle. When the rigid rock of the lithosphere bends and breaks, it makes earthquakes. Below the lithosphere, the rock is hot enough to flow and deform slowly, and thus does not produce earthquakes.

2. Here are the details explaining how we know about this 2.0-inch-per-year movement. In a classic paper, Tanya Atwater and Joann Stock (1998) used seafloor magnetic anomalies (magnetic stripes in the oceanic crust produced as the seafloor spreads and the Earth's magnetic field periodically reverses) to calculate precisely the motion of the Pacific Plate relative to North America over the last thirty million years. They determined that, from thirty million to twelve million years ago, the Pacific Plate traveled 1.3 inches per year in a direction of N60°W relative to North America. Twelve million years ago, the plate's speed picked up to 2.0 inches per year relative to North America. Eight million years ago, the plate's direction shifted a bit (to N37°W) but maintained a movement of 2.0 inches per year. Further evidence for this rate of motion comes from tracing the San Andreas fault south

into Mexico, where it opens into the Gulf of California. The Gulf began opening about five million years ago as Baja California tore away from North America as part of the Pacific Plate. Matching up rock formations in Baja and mainland Mexico that were connected until five million years ago, but which are now separated by 160 miles, confirms a movement rate of 2.0 inches per year (Oskin, Stock, and Martín-Barajas 2001).

3. To see continuously updated maps of earthquakes in California and Nevada during the latest hour, day, or week, check the U.S. Geological Survey's Earthquake Hazards Program website, which at this writing was located at http://earthquake.usgs.gov/earthquakes/recenteqscanv.

4. The Global Positioning System (GPS) uses a constellation of twenty-four U.S. military satellites that beam down precisely timed radio signals. A GPS receiver calculates the distance to each satellite by measuring how long the radio signals take. Signals received from several satellites simultaneously allow the receiver to triangulate its position to within a fraction of an inch. Repeating GPS measurements at the same location for several years in a row yields both the speed and direction of movement.

5. A misconception that I hear frequently is the idea that coastal California will sink into the sea if it splits away from the rest of the continent along the San Andreas fault. This is a physical impossibility. Continents are made of relatively thick, low-density rock that floats buoyantly in the denser rock of the mantle. This buoyancy—called isostatic equilibrium—keeps the continents above sea level in most places. A branch that breaks away from a floating log won't sink; its buoyancy doesn't depend on being attached to the log. A piece of a continent that breaks away from the rest won't sink into the mantle for the same reason.

PART II

THE BASIN AND RANGE
AND THE GREAT BASIN

When the traveler from California has crossed the Sierra and gone a little way down the eastern flank, the woods come to an end about as suddenly and completely as if, going westward, he had reached the ocean. From the very noblest forests of the world he emerges into free sunshine and dead alkaline lake levels. Mountains are seen beyond, rising in bewildering abundance, range beyond range. But however closely we have been accustomed to associate forests and mountains, these always present a singularly barren aspect, appearing gray and forbidding and shadeless, like heaps of ashes dumped from the blazing sky.

JOHN MUIR, 1918

Supreme over all is silence. Discounting the cry of the occasional bird, the wailing of a pack of coyotes, silence—a great spatial silence—is pure in the Basin and Range. It is a soundless immensity with mountains in it.

JOHN MCPHEE, 1998

Thus far it has been a labyrinth of mountains, irregular highlands, and frightful gorges—very interesting to the geologist and geographer, but dreadfully wearisome to the traveler, as we can vouch.

CORNELIA FERRIS, CROSSING THE GREAT BASIN ON HER WAY
TO CALIFORNIA IN 1853

6

WHERE RIVERS DIE

In May 1804, at the behest of President Thomas Jefferson, Meriwether Lewis and William Clark set out for the Pacific Ocean. Jefferson hoped the explorers would find a viable river trade route to the Pacific across the new lands of the Louisiana Purchase. Ideally, a short overland portage would link the headwaters of the Missouri River with those of the Columbia River. Such a connection would enrich the young nation with Pacific trade and establish an American presence in the Pacific Northwest.

The young captains came back two years later with good news and bad news. The expedition had crossed the Rockies and reached the Pacific. It had mapped sizable swaths of America's new western holdings. It had cataloged hundreds of new species of animals and plants and had established relations (mostly cordial) with dozens of resident Indian tribes. But it was disappointed in its primary objective. There was no navigable waterway to the Pacific—not even close. Between the upper reaches of the Missouri and Columbia rivers lay a vast heap of jagged mountains 200 miles across. Lewis and Clark had discovered the reality of the Rocky Mountains. Optimism about a possible river trade route to the Pacific faded.

But not for long. In 1812, fur trapper Robert Stewart stumbled across South Pass, a smooth divide across the Rockies in what is now southwestern Wyoming. The discovery of South Pass rekindled speculation about a possible trans-western river trade route. Beginning in the early 1700s, ship captains had reported seeing large rivers flowing into the Pacific along the northern California coast, and by the late 1770s explorers had discovered two big rivers emptying into the head of San Francisco Bay (the Sacramento

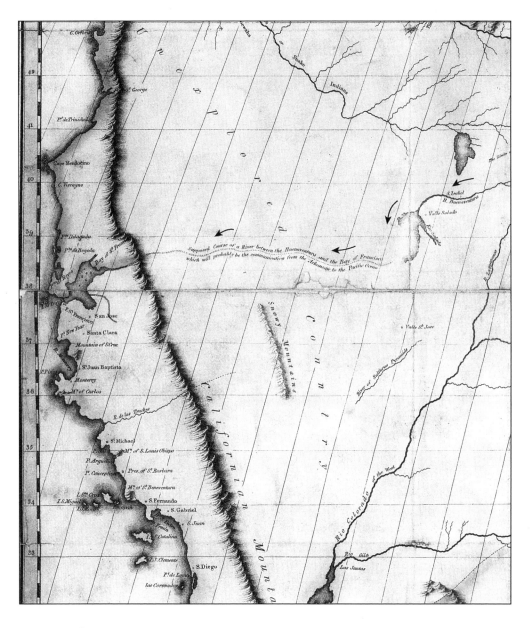

FIGURE 6.1

The mythical Buenaventura River appears on this 1816 map by Philadelphia mapmaker John Melish. The river flows west from the Rocky Mountains (out of view to the right) into a large lake (possibly Great Salt Lake) and then continues west to punch through the Sierra Nevada and flow into San Francisco Bay. The label along the river reads: *Supposed Course of a River between the Buenaventura and the Bay of San Francisco which will probably be the communication from the Arkansaw* [Arkansas River, which flows east out of the Colorado Rockies toward the Mississippi] *to the Pacific Ocean*. The "Unexplored Country" that the river bisects encompasses the then-unknown Great Basin. Arrows along the Buenaventura have been added for clarity.

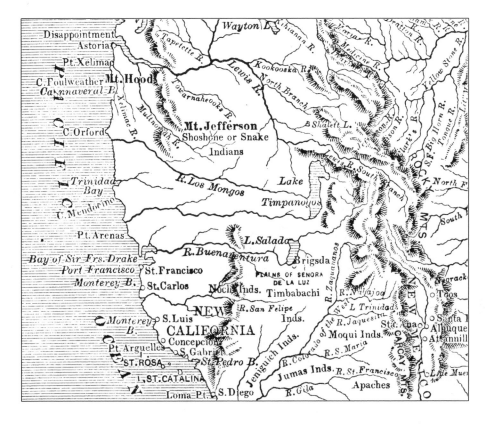

FIGURE 6.2

On this influential 1826 map by Albert Finley, no fewer than three rivers arising in the Rocky Mountains flow west to the California coast. The *R. Buenaventura* flows into San Francisco Bay, and the *R. Timpanogos* and the *R. Los Mongos* reach the coast north of the bay. To accommodate these rivers, Finley leaves a mountain-free gap between the north end of the Sierra Nevada and the south end of the Cascade Range. In 1844, the explorer John C. Frémont showed that the Sierra Nevada forms an unbroken barricade that traps interior rivers in the Great Basin.

and San Joaquin rivers). Did one or more of these rivers arise at the western base of the Rocky Mountains, perhaps near South Pass?

Enterprising minds began to churn with the financial opportunities that such a trans-western river would present. "During the early 19th century, a succession of explorers, trappers and traders—upon confronting a stretch of river they could not identify—rushed to the hopeful conclusion that they had stumbled upon the fabled river," historian Tom Chaffin writes. Rather magically, this trans-western river began to appear on early nineteenth century maps. It was called the Buenaventura. From the Rockies, it flowed west to California across the desert region where Nevada and Utah are today (figures 6.1 and 6.2). In the early nineteenth-century, this region formed a vast, unknown space that maps labeled "Unexplored Country" or "Terra Incognita." The Buenaventura

Border of the Great Basin

Border of the Basin and Range

FIGURE 6.3

A virtual satellite view looking obliquely north across the western United States. The dashed white line marks the borders of the Basin and Range Province, and the dotted white line encircles the Great Basin. Thin black lines mark state and international borders. Notice in the Basin and Range the washboard-like pattern of north–south-trending basins and ranges, produced by east–west stretching of the crust.

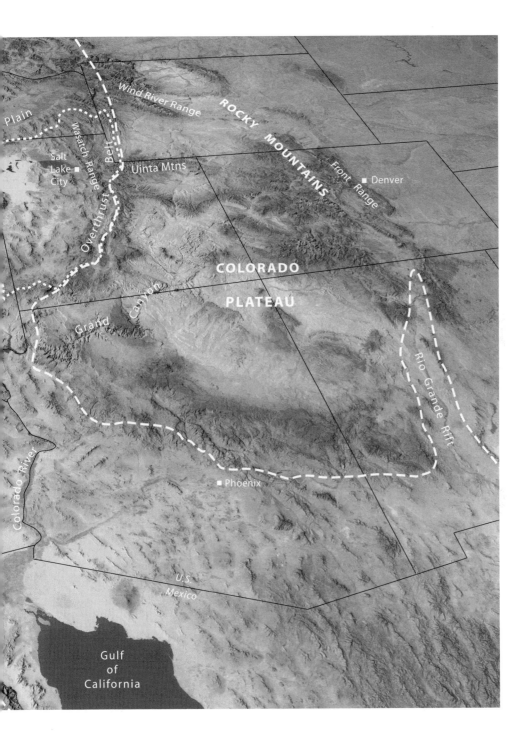

If you could slide the rocks of the Basin and Range Province back to their pre-stretching configuration, the Sierra Nevada would rise on the western outskirts of Las Vegas, and the Pacific Ocean would begin where the Sierra Nevada is now. The presence of the Great Basin makes impossible a Buenaventura-type river—that is, a river that arises in the Rocky Mountains and flows all the way to the California coast.

flowed across this space, in black ink that welled up from the headwaters of simple logic. The Rockies are high and snowy. Rivers begin there. Rivers flow downhill to the sea. Rivers may pause for a time in depressions to make lakes, but they soon flow out the low end of those depressions and continue toward the sea. At least one major river, and probably several, must flow west from the Rockies to the California coast. How could it be otherwise?

"What gets us into trouble is not what we don't know," Mark Twain once remarked, "it's what we know for sure that just ain't so." There is an alternative geography. Rivers can simply end. Lassoed by mountains, rivers may never reach the sea. Instead, aridity may suck them into the sky.

The Great Basin forms a vast, inward-draining desert bowl that occupies most of Nevada, half of Utah, and parts of four adjacent states (figure 6.3). The Sierra Nevada forms its western border, the Wasatch Range and the Colorado Plateau its eastern border. To the north and south, the Great Basin's boundaries are marked by low divides that separate streams that flow into the basin from those that flow away to link up, eventually, with the Snake River (to the north) or the Colorado River (to the south).

Great Basin rivers run downhill until they run out of downhill, ending in salty lakes (such as Utah's Great Salt Lake or Nevada's Pyramid Lake and Walker Lake) or evaporating from salt-encrusted depressions called sinks. No Buenaventura-like river can connect the Rockies to the California coast because no river can exit the Great Basin. The closest thing to the Buenaventura is Nevada's Humboldt River. The Humboldt arises in the mountains of northeastern Nevada and flows west-southwest for 350 miles across the state. Along the way, it fights a losing battle with the surrounding desert. With every mile, ground seepage and evaporation take more water from the river than its meager tributaries bring in. Some seventy miles east of the Sierra Nevada, the river gives up, pooling to a stop at the Humboldt Sink.

The Great Basin sits within a larger geologic province called the Basin and Range (figure 6.3). Nearly twice the size of Texas, the immense Basin and Range Province extends from the Sierra Nevada east to the Wasatch Range, and south from Oregon and Idaho far into Mexico. Its name comes from its regular tempo of north–south-trending mountain ranges separated by wide, gravel-filled basins. You can't go anywhere in the Basin and Range and be out of sight of mountains. The region contains some 500 ranges, most of them long (forty to one hundred miles), narrow (ten to twenty miles), and lined up north–south. Death Valley, California, is the lowest basin in the Basin and Range (282 feet below sea level), but for the most part, the Basin and Range is high country. Many of the basins in Nevada lie 4,000 to 7,000 feet above the sea, and some of the ranges reach over 13,000 feet.

To a degree unmatched anywhere on Earth, the Basin and Range is a stretched land. The Earth's crust here has stretched, east to west, like an accordion, by as much as 250 miles. This singular fact explains why the Great Basin lies *within* the Basin and Range.

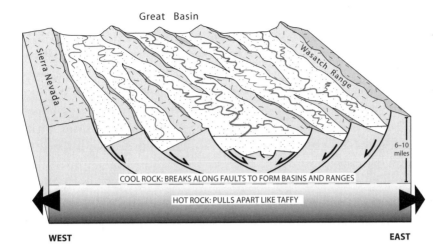

Great Basin

Sierra Nevada

Wasatch Range

6–10 miles

COOL ROCK: BREAKS ALONG FAULTS TO FORM BASINS AND RANGES

HOT ROCK: PULLS APART LIKE TAFFY

WEST

EAST

FIGURE 6.4

Throughout the Great Basin, east–west stretching has sundered the Earth's crust into north–south-trending mountain ranges flanked by broad, gravel-filled basins. The cool rock of the upper crust breaks into separate blocks along large, curving faults. The hot rock below it, stretching like taffy, acts like a rug pulled out from under the blocks above, rotating them along the curving faults. The up-rotated parts of the blocks form mountain ranges; the down-rotated parts form basins now filled with several thousand feet of rock debris eroded from the ranges. The diagram shows that west-tilting ranges dominate the western half of the Great Basin and east-tilting ranges dominate the eastern half, although there are plenty of exceptions to this rule. The diagram is schematic and vertically exaggerated; there are many more ranges between the Sierra Nevada and the Wasatch Range than I show here, and the ranges are much smaller relative to the full width of the Great Basin (as you can see on figure 6.3).

As the crust stretched and broke up into basins and ranges, it sagged in the center to form the inward-draining bowl of the Great Basin (figure 6.4). Although the term "Great Basin" suggests a smooth bowl with sloping sides, the Great Basin is no less rugged or mountainous than any other part of the Basin and Range.

This chapter tells two intertwined stories of the Great Basin: the human story of its discovery and the geologic story of this mountainous bowl where rivers die.

.   .   .

1843: Pioneers roll west in increasing numbers along the Oregon Trail, as the fever of westward expansion rises across the nation. The Buenaventura River still flows boldly across maps of the West, but the vast interior desert that it supposedly traverses is becoming less of a geographic blank. Explorers Jedediah Smith and Joseph Walker have passed through parts of this desert, and they haven't found the Buenaventura. The river is running out of places to hide.

The celebrated explorer John C. Frémont knew that the Sierra Nevada held the key to the Buenaventura River. If the river was real, it had to either cut through or go around the Sierra Nevada to reach the Pacific Ocean. In November 1843, Frémont set out from the Columbia River in Oregon, intent on finding the Buenaventura. He planned to ride south along the Cascade Range and the entire 400-mile-long Sierra Nevada. If the Buenaventura existed, his path would surely cross it. On "the best maps in my possession," he wrote, the Buenaventura formed a "connected water-line from the Rocky mountains to the Pacific ocean." Frémont was so confident about finding the river that he planned to overwinter on "the banks of the Buenaventura, where . . . our horses might find grass to sustain them, and ourselves be sheltered from the rigors of winter and from the inhospitable desert."

By the first week of January 1844, the Frémont expedition had reached the parched desert lands of what is now northwestern Nevada. "The appearance of the country was so forbidding that I was afraid to enter it," Frémont admitted. Yet he pressed on, "keeping close along the [eastern edge of the Sierra Nevada] mountains, in the full expectation of reaching the Buenaventura River." By mid-January, he was growing uneasy. He had reached the area where he thought he would find the river (the region a little south of where Interstate 80 drops out of the Sierra Nevada to cross the Great Basin today). "With every stream I now expected to see the great Buenaventura," he wrote anxiously, "and Carson [Kit Carson, Frémont's trusted guide] hurried eagerly to search, on every one we reached, for beaver cuttings, which he always maintained we should find only on waters that ran to the Pacific." Frémont discovered and mapped several good-sized rivers, including the Salmon Trout (today's Truckee), Carson, and Walker rivers. But they all flowed the wrong way—east out of the Sierra Nevada instead of west through it. Frémont hoped that one of them would prove to be a tributary of the Buenaventura still farther south. Instead, he found that they simply ended in salty lakes or sinks.[1] He began to suspect that the assumed geography of the West might be astoundingly wrong. Perhaps, he reasoned, the Sierra Nevada forms an unbroken wall that denies rivers access to the sea.

When Frémont, in his search for the Buenaventura River, rode south along the eastern face of the Sierra Nevada, he saw the mountains as we do today, rising like a series of giant steps from the western Nevada desert. The transition from desert to high alpine range crest takes place in just a few miles, yet climatically it mimics a trip from Las Vegas to Yellowstone. Sere slopes of grass and sagebrush meld upward into forests of juniper and pine, which give way, in turn, to ice-carved peaks. From the comfort of a modern highway, the scene is dramatic and lovely. From a nineteenth-century mule or wagon headed to California, it was awful to behold. For gold rush pioneer Elisha Perkins, it was "exactly like marching up to some immense wall built directly across our path." The Sierra Nevada's eastern escarpment is arguably the most striking piece of geography in the American West. What made it? The answer links directly to the formation of the Great Basin.

As we saw in chapter 3, some forty-five million years ago what is now the Sierra Nevada formed the western flank of the Nevadaplano, a two-mile-high, Altiplano-like highland that occupied all of Nevada and most of Utah—precisely where the Great Basin lies today. Rivers flowed west out of this highland, carpeting future California with the auriferous gravels. Starting about twenty million years ago, the crust of the Nevadaplano began to stretch east to west (for reasons that I'll explain shortly) and break up along a host of north–south-trending faults. The westernmost set of these faults uncoupled the Nevadaplano's western flank (the future Sierra Nevada) from the region to the east (the future Great Basin). The center of the Nevadaplano sagged like stretched pizza dough to become the Great Basin, while the Nevadaplano's western flank tilted upward like the rising end of a seesaw to become the Sierra Nevada.

The zone of faults that let this happen is known as the Eastern Sierra fault system—a band of active earthquake faults that parallels the Sierra's eastern escarpment. By any geologic definition, the faults of the Eastern Sierra system are active; any one of them could thrash Reno or Carson City with a violent earthquake any day. Over the last several million years, these faults have lifted the Sierra Nevada while lowering the Great Basin (top diagram in figure 6.5; see also the bottom diagram in figure 3.5).

The best place I know to lay your hands on one of these mighty faults is at the base of the Carson Range about fifteen miles south of Carson City. The Carson Range—which forms part of the Sierra Nevada's fractured eastern escarpment—rises abruptly 4,000 feet from the Carson Valley to the east. At its foot lies the Genoa fault, gloriously exposed in a small quarry not far from the town of Genoa (figure 6.5). Here—if you politely ask the locals to take a break from executing beer bottles with assorted firearms—you can touch one of the faults that made the Great Basin. The fault plane, which forms the back of the quarry wall, is scarred with parallel grooves scraped on the rock as the Carson Range rose and the Carson Valley dropped. I never tire of passing my hands over these gouges; it's like touching a tiger's back—I can almost sense the mountain gathering for another upward lunge. Mountains like the Carson Range and the Sierra Nevada don't rise gently. They jolt upward in spasms, a few inches or feet at a time, when their bounding faults shift during earthquakes. These periodic jerks have thoroughly pulverized the rock along the Genoa fault; it's so soft that I can dig out pieces with my fingers. But as I scan the shattered mess, I see something familiar: fragments of granite, ranging in size from marbles to golf balls—the rock of the high Sierra Nevada to the west. The once-solid granite here has been ground to bits, as if between millstones, by the lurching movements of the Genoa fault.

By the time Frémont passed by the Carson Range in late January 1844, he had grown ever more doubtful about the Buenaventura River. Reaching the east fork of the Walker River, he wrote, "We were sanguine to find here a branch of the Buenaventura, but were again disappointed." The absence of the Buenaventura put Frémont in a tough spot. It was now midwinter, and the expedition had traveled nearly 700 miles since leaving the Columbia River. Fifteen of its animals had died, and many of those remaining were worn out and lame. Worse, provisions had now run perilously low. Frémont had hoped

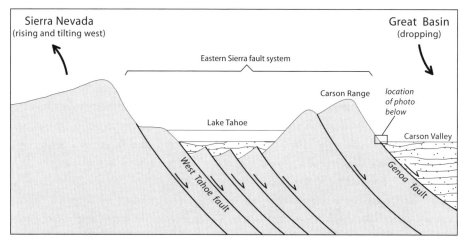

FIGURE 6.5

A zone of active earthquake faults called the Eastern Sierra fault system divides the Sierra Nevada from the Great Basin. Movements along this ten-mile-wide fault zone have raised the Sierra Nevada while dropping the Great Basin. Much of the activity has taken place during the last five million years, and it continues today. The result is steep and spectacular topography along the Sierra Nevada's eastern escarpment. The diagram (top) shows a cross-section of the Eastern Sierra fault system between the Sierra Nevada and the Carson Valley. The photograph (bottom) zooms in on the Genoa fault. The view looks north along the base of the Carson Range where the Genoa fault lies exposed in a small quarry south of the town of Genoa, Nevada.

to overwinter on the Buenaventura, anticipating abundant grass for his animals and game to feed his men. Now it appeared that he was stuck in a gigantic desert bowl formed by the unbroken escarpment of the Sierra Nevada. Faced with this turn of events, Frémont made a risky decision: to turn west and "cross the Sierra Nevada into the valley of the Sacramento, wherever a practicable pass could be found." Upon reaching Sutter's Fort in the Central Valley, the expedition would rest and resupply before continuing south on the final leg of its Sierra Nevada survey.

The notion of crossing the Sierra Nevada in *winter* seemed nothing less than suicidal to the resident Indians, to whom Frémont applied for the services of a guide. The natives "pointed to the snow on the mountain, and drew their hands across their necks, and raised them above their heads to show the depth; and signified that it was impossible for us to get through." But as Kit Carson later explained, "We were nearly out of provisions and cross the mountains we must, let the consequences be what they may."

The Indians told Frémont that it took "six sleeps," meaning six nights and seven days, to cross west over the Sierra Nevada in summer. It would take Frémont's expedition *six weeks,* from late January to early March 1844. Starting near present-day Bridgeport and heading for what is now Carson Pass, they broke trail through ever-deepening snows. Where the snow became impassably deep, they "were obliged to travel along steep hillsides, and over spurs, where the wind and sun had in places lessened the snow." The nights grew intensely cold, and the snow cut progress to a few backbreaking miles each day. "The people were unusually silent, for every man knew that our enterprise was hazardous, and the issue doubtful," Frémont recorded. Food became a pressing issue. "Our provisions are getting fearfully scant," and "all the men are becoming weak from insufficient food." Frémont ordered several horses and dogs slaughtered for meat.

Approaching Carson Pass, the snows became so deep that the horses sank up to their ears. Progress came to a standstill. The men cut mauls from trees to stamp down a trail over the snow and built snowshoes and sledges. They cut pine boughs and laid them across the compacted snow for extra traction and support. In this manner, they crept forward, building a road across the snow as they went. On February 14, Frémont saw "the dividing ridge of the Sierra"—meaning Carson Pass—a few miles ahead. It took six more days to get there. On February 20, the men clawed their way up the final slope to the 8,570-foot pass. Frémont was ecstatic. "We encamped with the animals and all the *materiel* of the camp, on the summit of the PASS on the dividing ridge," he triumphantly recorded.[2]

Calculating his elevation using the boiling point of water, Frémont concluded, to his amazement, that he was "2,000 feet higher than the South Pass in the Rocky mountains." Although he overestimated his elevation, he correctly surmised that he stood on "a range of mountains still higher than the great Rocky mountains themselves." The implications were clear. "This extraordinary fact accounts for the Great Basin, and shows that there must be a system of small lakes and rivers here scattered over a flat country, which the extended and lofty range of the Sierra Nevada *prevents from escaping to the Pacific ocean* [my italics]."

If the conceptual death of the Buenaventura River and the discovery of the Great Basin can be traced to a single moment, it was that one, when Frémont stood, shivering and exultant, on the dividing ridge of the Sierra Nevada on February 20, 1844.

We've seen *how* the Great Basin formed—from the stretching and collapse of the ancient Nevadaplano along a host of earthquake faults—and we visited the Genoa fault (figure 6.5) as a case study of how that happened. But *why* did it happen, and why is it still happening? For the Great Basin is a geologic work in progress, still stretching, collapsing, and rattling with earthquakes. GPS measurements show that Reno and Salt Lake City are getting farther apart by about a half-inch each year. That's close to the speed at which Iceland is stretching along the Mid-Atlantic Ridge. We know why Iceland is stretching: a mid-ocean ridge runs through it (see figure 5.1). Is the Great Basin stretching for the same reason, because a mid-ocean ridge runs through it?

Perhaps a better way to think about it is that the Great Basin sits on top of where a mid-ocean ridge *would be* if North America, migrating west, hadn't moved in to take its place. The convergence of North America with this mid-ocean ridge—the ridge that once divided the Pacific Plate from the Farallon Plate—has produced the stretching crust of the Basin and Range Province and the Great Basin within it. To make sense of this story, I need to set it within the larger context of how and why mountains have reared up across the American West. So for the next few pages, we'll pull back from the Great Basin and look at the bigger picture.

Some 200 million years ago, North America began to break away from the supercontinent Pangaea. As the westering continent widened the Atlantic, it narrowed the Pacific by sliding over and annihilating several ancient oceanic plates. The destruction of these plates (either by their collision with the continent's edge or their subduction beneath it) pushed up the North American Cordillera—the great belt of mountains that stretches from Alaska to Panama along North America's western edge. The most important of these old oceanic plates was the Farallon Plate, whose subduction, over roughly the last 140 million years, has built most of the mountainous topography of the Cordillera.

Figure 6.6 portrays this history in four time panels. Each panel represents a slice into the Earth between the Pacific Ocean west of California and the Colorado Rockies (in other words, a swath that passes roughly east to west through the center of figure 6.3). Panel 6.6a shows the scene about 110 to 85 million years ago, in mid-Cretaceous time, during an episode of mountain uplift known as the Sevier Orogeny (orogeny means "mountain building"). The Farallon Plate plunges east under North America. Chunks of ocean floor scrape off at the trench, joining previously accreted terranes to assemble the Franciscan Complex, which we visited at the Golden Gate in chapter 1. East of there, an arc of volcanoes erupts where the Sierra Nevada is now, while below, granite plutons glom together to form the Sierra Nevada Batholith, which we explored in chapter 4. East of the volcanic arc, where the Great Basin is today, lies a band of distorted, mountainous rock called a fold-and-thrust belt. Here, sideways compression—caused, perhaps, by the pressure of the Farallon Plate slowing North America's westward movement—crumples the continent internally, so that the rocks bunch up like the folds in a carpet shoved against a wall. Looking down on this mid-Cretaceous world, you see a Cordillera much narrower than today's. It looks a lot like the Andes—a skinny mountain belt hugging the continent's western edge. To the east, where Colorado and Wyoming are now, you see no Rocky Mountains. Instead, you see a low, lush, dinosaur-dotted plain that carries rivers east out of the fold-and-thrust belt toward a vast inland sea covering what will later become the Great Plains.

Jumping ahead to the time portrayed in panel 6.6b, we see a dramatic eastward shift in mountain upheaval. This episode, called the Laramide Orogeny, culminates in the uplift of the Rocky Mountains of Colorado and Wyoming. An oceanic plateau—a 500-mile-wide welt of thick basaltic rock on the Farallon Plate—has slid down the trench. The buoyancy of this oceanic plateau has floated the Farallon Plate upward in the mantle so that it now slides flat beneath the continent like a board shoved under a rug. The flat-sliding Farallon Plate squeezes the Rocky Mountains up from deep in the continental crust. (We'll explore this story, and the evidence for it, in chapter 12.) Meanwhile, to the west, mountains that had arisen across Nevada during the earlier Sevier Orogeny have been beveled down to a broad highland plateau—the Nevadaplano, the source of California's auriferous gravels (chapter 3).

Jumping forward in time to panel 6.6c, the Farallon Plate has restored itself to normal subduction. The plate is busy cranking out magma again—and on a terrifying scale. Volcanic eruptions open fire across the Nevadaplano. The eruptions advance through time in a northeast-to-southwest wave—a progression that may reflect how the Farallon Plate bent back down after the flat-subduction episode of panel 6.6b. You can see the legacy of this volcanic outburst throughout the mountains of the Great Basin, where thousands of vertical feet of volcanic ash, pumice, and lava bear witness to volcanic cataclysms far beyond the scale of human experience. We'll come back to this story in chapter 8, for as you'll see, the awesome volcanism of this time period spawned much of the metallic wealth that we find in the Great Basin today.

Panel 6.6d portrays the most recent—and still ongoing—mountain-building episode: the east–west stretching of the crust to make the Basin and Range Province and the Great Basin within it. The situation presents us with a reversal of the mountain-building forces we've seen so far. Previous upheavals arose mostly from sideways squeezing, along with prodigious outpourings of volcanic lava and ash. Sideways squeezing happens wherever tectonic plates converge, and it is responsible for making most of the Earth's big mountain belts. (With the thumb and forefinger of one hand, squeeze

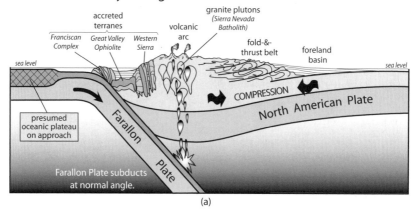

~ 110 – 85 million years ago

accreted terranes

Franciscan Complex | Great Valley Ophiolite | Western Sierra

volcanic arc

granite plutons
(Sierra Nevada Batholith)

fold-&-thrust belt

foreland basin

sea level

sea level

COMPRESSION

North American Plate

presumed oceanic plateau on approach

Farallon Plate

Farallon Plate subducts at normal angle.

(a)

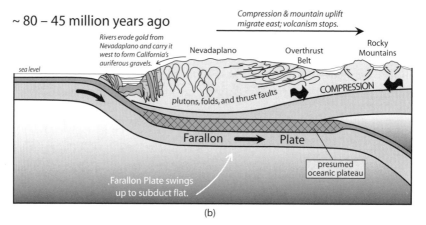

~ 80 – 45 million years ago

Compression & mountain uplift migrate east; volcanism stops.

Rivers erode gold from Nevadaplano and carry it west to form California's auriferous gravels.

Nevadaplano

Overthrust Belt

Rocky Mountains

sea level

COMPRESSION

plutons, folds, and thrust faults

Farallon ➡ Plate

presumed oceanic plateau

Farallon Plate swings up to subduct flat.

(b)

FIGURE 6.6

The plate tectonic history of the North American Cordillera in the western United States during the last 110 million years. The formation of the Basin and Range Province and the Great Basin are the latest events in a long history of mountain upheaval linked to the subduction and eventual annihilation of the Farallon Plate.

together the skin on the top of your other hand. The folded ridges on your skin look a lot like the Appalachians or the Hindu Kush because the mountains formed in a similar way. This trick, sadly, works better the older you get.) Notice, however, that during the time portrayed in panel 6.6d mountain-building forces switch from *squeezing* to *stretching*. Both forces—squeezing and stretching—can make mountains, albeit in different ways (figure 6.7).

We think that the switch from squeezing to stretching happened because of the extinction of the Farallon Plate. Going back to panel 6.6c, notice that, as volcanoes

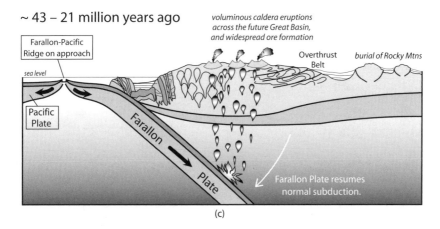

~ 43 – 21 million years ago

voluminous caldera eruptions across the future Great Basin, and widespread ore formation

Farallon-Pacific Ridge on approach

sea level

Pacific Plate

Farallon

Plate

Overthrust Belt

burial of Rocky Mtns

Farallon Plate resumes normal subduction.

(c)

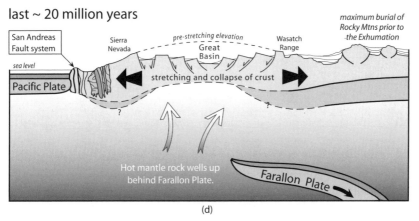

last ~ 20 million years

maximum burial of Rocky Mtns prior to the Exhumation

San Andreas Fault system

sea level

Pacific Plate

Sierra Nevada

pre-stretching elevation

Great Basin

Wasatch Range

stretching and collapse of crust

?

?

Hot mantle rock wells up behind Farallon Plate.

Farallon Plate

(d)

explode across the future Great Basin, North America is closing in on the Farallon-Pacific Ridge—the mid-ocean ridge that divides the Farallon Plate from the Pacific Plate. The continent's westward migration will carry it *across* that mid-ocean ridge, taking us forward in time to panel 6.6d. There the Farallon Plate, cut off from its source of seafloor spreading at the ridge, founders like a sinking ship into the mantle below North America. (You can see the dead plate down there today. It shows up on seismic images, which use earthquake waves to take pictures of the mantle somewhat like using X-rays to image your bones. I'll give you more on this story in chapter 13.) Under the Basin and Range, a large pocket of hot mantle rock wells up through the gap opened by the departed Farallon Plate. The result is east–west stretching of the crust to form the Basin and Range Province and its collapse to make the Great Basin— processes that continue today. Figure 6.8 portrays this story in map view. It shows how, as North America made contact with the Farallon-Pacific Ridge, it gave rise both to the San Andreas fault and the crustal stretching that made the Basin and Range and the Great Basin.

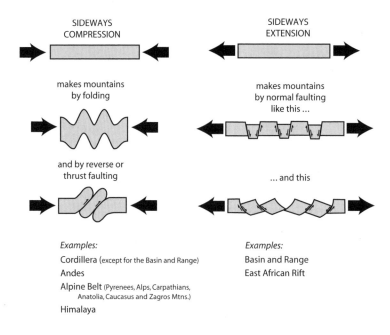

SIDEWAYS
COMPRESSION

makes mountains
by folding

and by reverse or
thrust faulting

SIDEWAYS
EXTENSION

makes mountains
by normal faulting
like this ...

... and this

Examples:
Cordillera (except for the Basin and Range)
Andes
Alpine Belt (Pyrenees, Alps, Carpathians,
     Anatolia, Caucasus and Zagros Mtns.)
Himalaya

Examples:
Basin and Range
East African Rift

FIGURE 6.7

Most of the Earth's major mountain belts arise from sideways squeezing or compression, which typically happens where oceanic plates subduct beneath continents or where two continents approach one another and collide. Sideways stretching or extension, which happens where plates pull apart, makes mountains like those in the Basin and Range.

By the time Frémont had crested the Sierra Nevada at Carson Pass in February 1844, he was almost certain that the Sierra "prevents from escaping to the Pacific ocean" any river coming from the east.[3] His mind had conceived a novel geography. East of the Sierra Nevada, he suspected, lay a great desert basin where rivers cannot reach sea.

To confirm his suspicions, Frémont needed to finish his Sierra Nevada survey. Heading down from Carson Pass toward Sutter's Fort in the Central Valley, Frémont and his men fought their way through "deep fields of snow" across "a large intervening space of rough-looking mountains." It took two weeks of toil through the snow-clogged forests before they finally staggered into Sutter's Fort, bone-weary and near starvation. But they had beaten the Sierra Nevada, in winter, without losing a man. On March 24, 1844, fully rested and resupplied, the expedition set out south along the Central Valley. Riding through the bloom of spring, the men looked east to admire the Sierra's "rocky and snowy peaks where lately we had suffered so much." Snowmelt from those peaks flowed into west-draining rivers, all of which arose high on the Sierra's western slope, and none of which passed *through* the range from the other side.

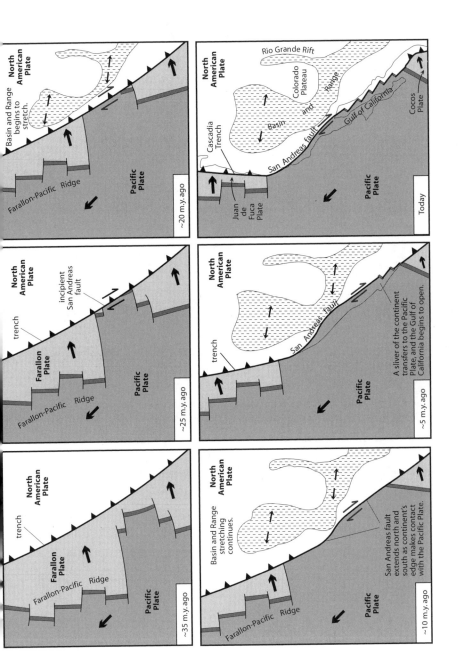

FIGURE 6.8

A map-view explanation of how the demise of the Farallon Plate produced the Basin and Range Province. Before about thirty million years ago, a continuous mid-ocean ridge—the Farallon-Pacific Ridge—separated the Pacific Plate from the Farallon Plate. From this ridge, the Farallon Plate spread eastward before subducting under the North American Plate. Starting about twenty-eight million years ago, the North American Plate began to intersect the Farallon-Pacific Ridge and make contact with the Pacific Plate on the other side. Because the North American Plate was heading west while the Pacific Plate was heading northwest, the result at their touching edges was not subduction but side-by-side sliding. A new type of plate boundary was thus born—a side-by-side sliding boundary that grew, north and south, to become the San Andreas fault and its southern extension, the Gulf of California. Two remnants of the Farallon Plate still subduct along North America's western edge, where the continent has not yet overridden the ridge. North of the San Andreas fault, a small Farallon Plate remnant called the Juan de Fuca Plate plunges down the Cascadia Trench to feed magma to the Cascade Range of volcanoes. South of the Gulf of California, a larger Farallon Plate remnant called the Cocos Plate dives under Mexico and Central America at the Central America Trench.

On April 14, the expedition rounded the south end of the Sierra Nevada at Tehachapi Pass. Six months after he had set out to find it, Frémont finally had his answer about the Buenaventura River:

It had been constantly represented [on many maps] . . . that the bay of San Francisco opened far into the interior by some river coming down from the base of the Rocky mountains, and upon which supposed stream the name of Rio Buenaventura had been bestowed. Our observations of the Sierra Nevada . . . show that this neither is nor can be the case. No river from the interior does, or can, cross the Sierra Nevada.

Having showed that the Buenaventura was a myth, Frémont announced his signal contribution to the geography of the American West:

The Great Basin—a term which I apply to the intermediate region between the Rocky Mountains and the next range [Sierra Nevada], containing many lakes, with their own system of rivers and creeks, (of which the Great Salt is the principal,) and which have no connection with the ocean, or the great rivers that flow to it.

NOTES

1. The Truckee River flows into Pyramid Lake, the Carson River ends at the Carson Sink, and the Walker River terminates in Walker Lake.

2. Five years after Frémont crossed it, Carson Pass became the main gateway through which gold rush emigrants entered California. "Had we met such an ascent in the earlier part of the journey," forty-niner William Kelly wrote of the climb to Carson Pass, "I fancy we should have pronounced it insurmountable, and turned back in despair." But there would be no turning back now—not this close to California. Up they went, using block-and-tackle strapped to trees, multiple teams of oxen, horses, or mules on each wagon, and all hands pushing and turning the wagon wheels.

3. The Middle Fork and North Fork of the Feather River are exceptions to Frémont's observation. They head in the Modoc Plateau in the far northeastern corner of California and cross the low northern end of the Sierra Nevada to reach the Sacramento River, and thus the Pacific.

THE GROWING PAINS OF
MOUNTAINS

Daily it is forced home on the mind of a geologist, that nothing, not even the wind
that blows, is so unstable as the level of the crust of this Earth.

CHARLES DARWIN, AFTER WATCHING THE ANDES LEAP HIGHER
DURING THE 1835 CHILE EARTHQUAKE

On a bright July afternoon some years ago, I found myself standing with dozens of
other people on a peak that, in summer, is one of the most crowded in the American
West. To reach Mount Whitney, you first do a lot of waiting. You wait in line for a hiking
permit at the ranger station in the tiny town of Lone Pine, two vertical miles below
Mount Whitney in the Owens Valley. Then you wait behind cars jockeying for limited
parking at the Whitney Portal trailhead. Then you wait on the trail, as day-hikers, back-
packers, and mule trains jam up periodically like cars on a crowded freeway. Mostly,
though (unless you're in excellent physical condition), you wait to gasp for air. There's
more heavy breathing on the trail to Mount Whitney than in most X-rated movies. But
if you grit it out, you eventually find yourself standing on an ice-carved granite knoll
14,505 feet above the sea: the highest point in the lower forty-eight U.S. states.[1]

Mount Whitney is like the tallest kid in your high school class—just the tallest of a
very tall group. Nine other Sierran peaks rise above 14,000 feet, most of them within
sight of Mount Whitney. In the coterminous United States west of the Rockies, 66 peaks
stand above 14,000 feet, and 844 above 13,000 feet. (By contrast, east of the Rockies only
five peaks stand higher than 6,000 feet.) Moreover, many of these western peaks are
still rising. Several of them, I'm certain, will be higher when you read this than they were
when I wrote it. The West is a land of living mountains—and earthquakes are their growing pains.

It was a typical day on Mount Whitney: bright sun, cobalt sky, and wind howling
out of the west. Squeezed upward by the Sierra Nevada's broad western ramp, the wind

Mt. Whitney
12 miles

Alabama Hills
(Sierra Nevada
foothills)

scarp of 1872 quake

SIERRA NEVADA BLOCK
rose six feet and slid north
(away from you) 15 feet
during the 1872 quake.

FIGURE 7.1

View north-northwest along the scarp of the Lone Pine fault one-half mile west of Lone Pine, California. This scarp grew six feet during the great 1872 earthquake, adding to the offsets of several previous quakes to give it its present height of fifteen to twenty feet. Distinctive brown rocks eroded from the nearby Alabama Hills begin on the east side of the scarp in the area marked by the arrow.

accelerated over the summit like air over a jet's wing. We hikers spread our arms like scarecrows and leaned backward against the wind, seeing how far we could go before falling over. To the east, the Basin and Range Province sprawled beyond our vision, its serial mountain ridges lined up to the horizon. Nowhere in the world is the story of earthquake-driven mountain uplift more vividly laid out than in the eastern Sierra Nevada and the Basin and Range. Earthquakes crackle daily across this broken land where young mountains are leaping skyward—right now, in human time.

To find these same rocks on the west side of the scarp, you need to climb up the scarp *and* step forty feet northwest (away from you in the photograph). The northwest offset shows us that the Sierra Nevada not only grows higher with each earthquake, it also leaps northwest. In the 1872 quake, the range rose six feet and slid northwest fifteen feet. Mount Whitney lies out of view about twelve miles to the west (left).

You don't need to climb Mount Whitney to see evidence of the forces that have jacked the Sierra Nevada two miles above the Owens Valley. You only have to go a half-mile west of Lone Pine, to the site shown in figure 7.1. It shows a fault scarp—a low cliff formed where a fault shifts. The scarp represents the surface rupture of the Lone Pine fault, one of dozens of faults that divide the Sierra Nevada from the Basin and Range.[2] At 2:25 a.m. on March 26, 1872, the Lone Pine fault ripped loose along this scarp, and in a few terrifying seconds the entire Sierra Nevada lurched upward six feet and slid northwest fifteen feet. Seismic waves thrashed the Owens Valley like a snapped bed sheet, turning Lone Pine into rubble and killing twenty-seven residents. Traveling outward like ripples on a pond, the seismic waves shook people awake throughout California and Nevada, including John Muir, asleep 110 miles away in his Yosemite Valley cabin:

> I ran out of my cabin, both glad and frightened, shouting, "A noble earthquake! A noble earthquake!" feeling sure I was going to learn something. The shocks were so violent and varied, and succeeded one another so closely, that I had to balance myself carefully in walking as if on the deck of a ship among waves. . . . The shocks became more and more violent—flashing horizontal thrusts mixed with a few twists and battering, explosive, upheaving jolts,—as if Nature were wrecking her Yosemite temple, and getting ready to build a still better one.

In the moonlight, Muir watched as great avalanches of rock fell to the floor of the Yosemite Valley, spewing friction sparks in "an arc of glowing, passionate fire." The aftershocks lasted for more than two months, and Muir, ever the student of Nature, "kept a bucket of water on my table to learn what I could of the movements," qualifying him, perhaps, as California's first seismologist. The 1872 Lone Pine earthquake was one of the three most violent quakes in recorded California history, comparable to the 1857 Fort Tejon and 1906 San Francisco quakes that I described in chapter 5. When the Lone Pine fault lurches again—as it most assuredly will—Mount Whitney will leap several feet higher, and the residents of Lone Pine will again pay the price.

·   ·   ·

Most of the Earth's mountains rise where faults shift. Faults, in turn, occur where the Earth's crust breaks in response to stress. Most often, those stresses arise from the movements of plates. The Earth's biggest mountain belts arise where plates converge, with one plate either sliding beneath the other (such as where the Juan de Fuca Plate is sliding under North America to make the Cascade Range) or crunching sideways into the other (such as where the Indo-Australian Plate is jamming northward into Asia to push up the Himalayas). In both cases, the Earth's crust is squeezed sideways and mountains lurch upward, earthquake by earthquake, in response. But mountains also arise from sideways-stretching forces where tectonic plates are tearing themselves apart (see figure 6.7). Today, sideways-stretching forces are actively hoisting mountains in East Africa, where the African Plate is splitting along the East African Rift Valley (see frontispiece figure), and in the Basin and Range.

Scan a geologic map of the Basin and Range, and you'll see a repeating east–west pattern: basin, fault, range; basin, fault, range, all lined up more or less north to south. Faults flank every range on one side, and sometimes both sides. Whenever the east–west-stretching forces of the region overcome the friction that locks one of these faults, the fault snaps and the rocks on each side leap to a new position. The range on one side of the fault jerks up a bit, the basin on the other side drops down, and the energy released surges outward as seismic waves. Because the Basin and Range is stretching, the basins and ranges also shift incrementally *apart* during these events.

Most Basin and Range earthquakes are too tiny for you or me to feel. Only seismographs (sensitive ground-motion instruments) pick them up. Nonetheless, these small quakes are so common (often as many as several dozen per day) that they contribute measurably to the stretching of the Basin and Range and the uplift of its mountains. Less frequent but more dramatic are the bigger leaps that the mountains take during larger quakes. After the largest quakes—those involving several feet of mountain uplift—fault scarps several feet high and several tens of miles long often mark the places where the range-bounding faults have snapped and shifted.

Although the entire Basin and Range is earthquake country, the quakes aren't distributed evenly across the region. In the southwestern United States, most earthquakes cluster into one of the four seismic belts shown on figure 7.2: the San Andreas belt, Walker Lane belt, Central Nevada belt, and Intermountain belt. The latter three belts occur partly or entirely within the Basin and Range. The overriding force triggering the quakes in all four belts seems to be, in large part, the relentless northwestward pull of the Pacific Plate. As we saw in chapter 5, the Pacific Plate, cruising northwest at about two inches per year, is dragging large sections of the southwestern United States along with it (see figure 5.6). You can see the influence of the Pacific Plate on many Basin and Range faults. For instance, recall that when the Lone Pine fault broke during the 1872 Owens Valley earthquake, it didn't just lift the Sierra Nevada *up*; the entire range also lurched *northwest*. That's a common pattern, particularly in the western Basin and Range. When a fault snaps, the rocks on the west side often leap northwest (in addition to shifting either up or down) as they yield to the Pacific Plate's northwestward tugging. But that tugging isn't being accommodated evenly; the Basin and Range isn't stretching uniformly like a rubber sheet. Instead, most of the stretching is taking place within those four seismic belts. You can see this in the graph at the bottom of figure 7.2. It gives the results of precision GPS measurements taken over several years at places along the east–west line shown on the map. The graph shows how fast areas along that east–west line are pulling northwest away from the rest of North America. Notice that the jumps on the graph occur across the seismic belts, showing us that the stretching of the Basin and Range is taking place mostly within the seismic belts. This makes sense—after all, earthquakes are what allow the Basin and Range to stretch. If the stretching that we see today stays focused on those seismic belts for, say, the next few million years, we can well imagine what the future will bring: continental breakup, perhaps along the lines of one of the scenarios that I posited in figure 5.7.

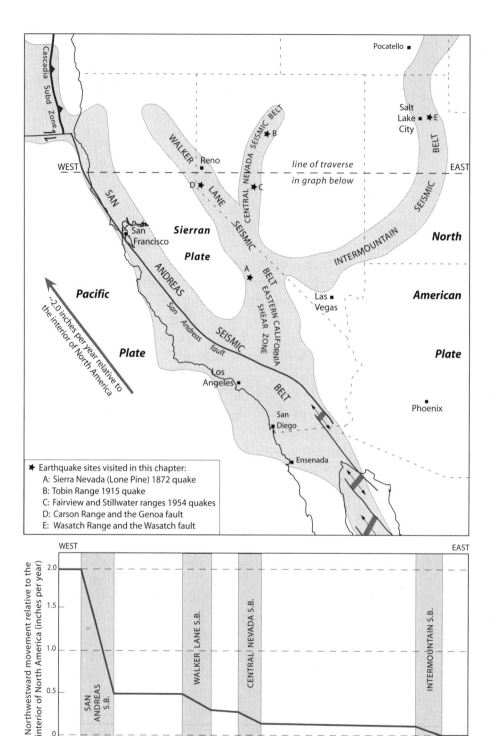

Pocatello ■

Salt
Lake ■ ✶ E
City

WEST — line of traverse — EAST
in graph below

CASCADIA SUBD. ZONE

WALKER LANE SEISMIC BELT

CENTRAL NEVADA SEISMIC BELT

INTERMOUNTAIN SEISMIC BELT

Reno ■
D ✶
✶ C
B ✶

SAN ANDREAS

SEISMIC

San Andreas fault

Sierran

Plate

Pacific

~2.0 inches per year relative to the interior of North America

Plate

San
Francisco ■

EASTERN CALIFORNIA SHEAR ZONE

A ✶

BELT

Las ■
Vegas

Los
Angeles ■

San
Diego ■

Ensenada ■

North

American

Plate

Phoenix ■

✶ Earthquake sites visited in this chapter:
  A: Sierra Nevada (Lone Pine) 1872 quake
  B: Tobin Range 1915 quake
  C: Fairview and Stillwater ranges 1954 quakes
  D: Carson Range and the Genoa fault
  E: Wasatch Range and the Wasatch fault

WEST   EAST

Northwestward movement relative to the interior of North America (inches per year)

2.0
1.5
1.0
0.5
0

SAN ANDREAS S.B.
WALKER LANE S.B.
CENTRAL NEVADA S.B.
INTERMOUNTAIN S.B.

The thing that dazzles me most about the data in figure 7.2 is that technology lets us track, in human time, the stretching of the Basin and Range and the uplift of its mountains. We can plot the earthquakes as they pop off day by day, we can see how those quakes cluster into seismic belts, and we can measure the stretching of the land and the uplift of the mountains from year to year.[3] In the Basin and Range, we are witness to the ongoing birth of mountains. Like few other places on Earth, this breaking land lifts the blinders of human time to give us a direct glimpse of geologic creation.

Shifting between human time and geologic time never comes easily, even to veterans of the profession. The mental leap is usually just too great. Yet the uplift of mountains, when it happens, takes place at the scale of human time. A fault shifts. Seismic waves thrash outward. The Earth's crust rips for miles, and quadrillions of tons of rock suddenly leap higher into the sky.

And then, nothing—or so it often seems. After a mountain jumps during an earthquake, centuries may pass before it jumps again. What goes on during these in-between times?

Earthquakes lift mountains up, but erosion takes them down. In human time, erosion acts with near-imperceptible slowness (floods and landslides being notable exceptions). Grain by grain, wind, rain, rivers, and ice chip away at rock. Compare a century-old photograph of a mountain with the same view today, and you'll rarely see much evidence for erosion. In human time, mountains take on an illusion of permanence. Throughout literature and culture, they symbolize all that is steadfast, unyielding, adamantine. But mountains in geologic time are little more than temporary bumps, able to grow only where rates of uplift outpace rates of erosion. In the race between the leaping hare of uplift and the grinding tortoise of erosion, erosion always wins eventually. The forces of uplift can lead the race for millions of years, but they can't outlast erosion. As the Earth's tectonic plates shift into new configurations, the forces of uplift relocate, abandoning mountains to the incessant gnawing of erosion.

Throughout much of the American West, the hare of uplift leads the tortoise of erosion as earthquakes push mountains toward the clouds faster than the clouds can take

FIGURE 7.2 (OPPOSITE)
Most earthquakes in the southwestern United States cluster along the four seismic belts labeled on the map (top). Locations A through E mark earthquake sites that we visit in this chapter. Note the east–west line of traverse, which relates to the graph below the map. The graph (bottom) shows how fast (in inches per year) various points along that east–west line are moving northwest relative to the interior of North America, based on precision GPS measurements. The jumps on the graph represent increased rates of movement toward the northwest. Notice that these jumps occur within the seismic belts, telling us that most of the rifting of the Basin and Range is presently happening in those seismic belts. The main force behind this rifting appears to be the northwestward pull of the Pacific Plate.

them down. But the West's present topography is recent. Most of it has risen and roughened within the last one hundred million years, and much of it much more recently than that. This makes the West as youthful, relative to the Earth itself, as a two-year-old toddler compared to her ninety-year old great-grandmother. Had you visited North America 250 million years ago, you would have seen a mirror-image reversal of continental topography. The West then was mostly smooth and low lying—much like the East today—whereas in the East, Himalayan-sized mountains formed a colossal barricade stretching from Nova Scotia to Georgia. These American Himalayas arose when North America clunked into Eurasia and Africa in the process of assembling the supercontinent Pangaea. Later, when North America split away from Pangaea and headed west, the forces of uplift relocated to the continent's western edge. Mountains began to nose skyward from Alaska to Central America, while erosion chewed the American Himalayas down to the nubs that we call the Appalachians. (Throughout the bedrock of the Appalachians, you can find minerals that form only under the pressure of several miles of rock overhead, testifying to the weight of great mountains since removed.) Some years ago, I took friends from the East on a tour of the backcountry near my San Diego home. As we passed a range of pretty hills, they asked, "What are those *mountains* called?" Our map showed that these bumps had no name, even though they stand higher than 90 percent of the United States east of the Rockies. (We dubbed them the "Sierra sin Nombre" in honor of their namelessness.) West of the Rockies, the standard for what constitutes a namable mountain rises higher.

.   .   .

10:53 pm, October 3, 1915, Central Nevada Seismic Belt: Pleasant Valley lies dark and still between the black silhouettes of its flanking ranges. Suddenly, with a roar like an artillery barrage, the Tobin Range east of the valley rips free along its bounding fault. A fresh scarp rifles across the foothills as the Tobin Range jumps eighteen feet. Ranch houses slew sideways off their foundations. Walls crack and crumble. Chickens and pigs are crushed in collapsing coops and pens. Water tanks pitch over in the towns of Battle Mountain and Lovelock, fifty miles away. People are rattled out of bed as far away as Salt Lake City, Portland, and Los Angeles. Springs that had flowed reliably for years dry up and fresh ones burst forth elsewhere as groundwaters find new routes through the fractured crust. The scarp of this 1915 quake still forms a bold scar along the west flank of the Tobin Range (figure 7.3).

3:07 a.m., December 16, 1954, Central Nevada Seismic Belt: The roar of splitting rock shreds the night as Fairview Peak lurches up seven feet and slides northwest eleven feet. Four minutes and twenty seconds later, another quake—probably triggered by stresses released from the first—rips loose in the Dixie Valley twenty miles to the north. Seismic waves thrash the desert, and within seconds the Stillwater Range lurches up nine feet. The first motorist along nearby Highway 50 the next morning

Tobin Range

scarp of 1915 quake

FIGURE 7.3

The violent uplift of Basin and Range mountains during earthquakes leaves scars, in the form of fault scarps, that mark the surface ruptures of the range-bounding faults. The photograph looks east from Pleasant Valley toward the Tobin Range and the scarp of the October 3, 1915, earthquake, which raised the Tobin Range eighteen feet that day. Tens of thousands of such lurchings have built the mountains of the Basin and Range.

comes to a screeching halt; the road has been ripped into a series of ledges by the fresh scarp of the Fairview Peak fault. Shaking is reported throughout Nevada, California, Arizona, Utah, Idaho, and Oregon. Today, the scarps from these twin 1954 quakes wind for sixty discontinuous miles along the flanks of the Fairview and Stillwater ranges. By beaming sound waves into the ground and tracking how those waves echo back, geologists have mapped out the underground trace of the Dixie Valley fault—the one that pushed the Stillwater Range nine feet higher that day (figure 7.4).

4:30 p.m., June 2, 2008, east face of the Carson Range, Walker Lane Seismic Belt: There's no earthquake—at the moment. I'm revisiting the Genoa fault, which, as you may remember from chapter 6, is the one that hoisted western Nevada's Carson Range (see figure 6.5). There haven't been any catastrophic earthquakes on the Genoa fault since pioneers settled the Carson Valley more than 150 years ago. But 150 years of quietude doesn't mean much; an active fault may sit still for centuries before jacking up a mountain again. I'll use the Genoa fault as a case study in how we can judge a fault's activity over a geologically meaningful time span—an important step in assessing the earthquake risk to a region. The Genoa fault hasn't ripped loose with a huge quake lately, but all evidence indicates that it is an earthquake time bomb. Two things testify to this: features of the Carson Range that indicate

WEST   EAST

WEST   Horizontal distance in feet   EAST

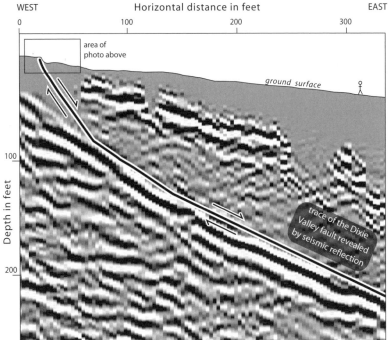

FIGURE 7.4

The photograph (top) looks north along the scarp formed by the December 1954 Dixie Valley earthquake. Fault scarps represent the surface ruptures of faults that penetrate miles into the Earth's crust. The only way we can see these faults belowground is to use seismic reflection, a technique similar to imaging a person's innards using ultrasound. Artificial seismic waves (created by controlled explosions or by large ground-stamping hydraulic hammers) are beamed into the ground. Where these waves hit changes in the rock (like along a fault), they echo back to give us an underground side view, or profile. The seismic reflection image (bottom) shows the trace of the Dixie Valley fault, which raised the Stillwater Range nine feet during the 1954 earthquake.

recent, rapid, and ongoing uplift, and evidence of prehistoric earthquakes along the fault.

Mountains, like faces, reveal their history in their creases and lines. A mountain that is jolting upward rapidly along a fault looks different from one that is being pushed up slowly or not at all (figure 7.5). The first clue comes from the straightness of the range front. Fast-rising ranges typically have straight range fronts, whereas slow-rising ranges have sinuous range fronts made uneven by erosion. A second clue comes from the canyons that crease the range front: short, steep canyons go with fast-rising ranges; long, deep canyons go with slow rising ranges because erosion has had more time to cut them. A third clue comes from the size of the alluvial fans—those sloping layers of gravelly debris eroded from the range and washed out into the neighboring basin during rainstorms. Other things being equal, a fast-rising range will have tiny alluvial fans, whereas a slow-rising range will have larger fans because erosion has had more time to carry rock debris from the range out into the basin. A fourth clue comes from the shapes of the range's ridges where they terminate along the range front. In a fast-rising range, the ridges often terminate in large triangular facets where their ends have been nipped off by uplift along the range-bounding fault. If the fault hasn't moved in a while, erosion will have beveled off and downsized these facets.

If you compare figures 7.5 and 7.6, you can see that the Carson Range (along with Utah's Wasatch Range, which we'll explore next) passes every test for a fast-rising mountain. The range front runs straight. It has short, steep canyons. Tiny alluvial fans spill from those canyons. And its ridges end in large triangular facets above the Genoa fault. Conclusion: the Genoa fault's recent quietude is, in all likelihood, just a pause between violent upward lunges. And that's not good news. In the last century, human settlement has exploded across certain sectors of the West. Today, more than a half-million people live within a few miles of the Genoa fault in the greater Carson City, Reno, and Lake Tahoe areas. Every feature of the Carson Range and the Genoa fault announces "big earthquake coming!" But when? And when it does come, how bad will it be?

To find some answers, geologists at the University of Nevada at Reno have dug trenches across the Genoa fault and age dated (using radioactive carbon) the remains of plant debris split by the fault during past earthquakes. They have found evidence for two major past quakes: one about 2,100 years ago and another about 550 years ago. Each quake hoisted the Carson Range an estimated *ten to eighteen feet*! If these findings reflect typical behavior on the Genoa fault, they tell us two things. First, the fault doesn't snap very often—many centuries may pass between big quakes. But *when* it snaps, it does so in a big way. Ten to eighteen feet of displacement is a calamitous earthquake by any measure. If you doubt it, look at Tables 7.1A and 7.1B, which list major historic and prehistoric Basin and Range earthquakes by their displacements, scarp lengths, and Richter magnitudes. (See footnote 2 in Table 7.1A for an explanation of Richter magnitude.) The comparison shows that the Genoa fault (along with Utah's Wasatch fault, which we'll

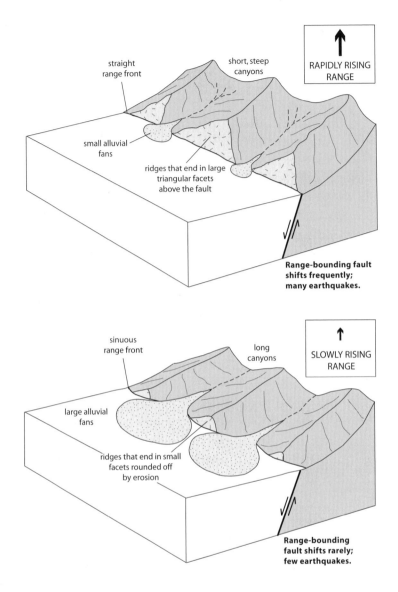

FIGURE 7.5

How to tell how fast a mountain is rising along its range-bounding fault. Compare these diagrams with the photographs of the Carson and Wasatch ranges in figure 7.6.

explore next) is capable of firing off quakes comparable in magnitude to the largest quakes ever recorded historically in the Basin and Range.

So far, even the most violent Basin and Range earthquakes have produced few fatalities and little damage because they have occurred in sparsely settled regions. But as urban development metastasizes near active faults, the potential for tragedy escalates. The active faults of the Basin and Range represent hundreds of earthquake time bombs, and there's no way to tell where or when the next one will go off. The Carson City–

FIGURE 7.6

Examples of two rapidly rising, earthquake-prone Basin and Range mountains, both of which lie near areas of dense urban development. The top photograph looks northwest across the Carson Valley toward the eastern face of Nevada's Carson Range, uplifted along the Genoa fault. The bottom photograph looks east toward Utah's Wasatch Range, uplifted along the Wasatch fault. (Photograph by Rod Millar, Utah Geological Survey, taken from the air over Mapleton a few miles south of Provo.) Comparing these images to figure 7.5, you can see that both ranges qualify as fast-rising mountains bounded by active faults.

TABLE 7.1A Selected Historic Earthquakes in the Basin and Range

Historic Quakes[1]	Vertical displacement	Horizontal displacement	Scarp length	Richter magnitude[2]
1872 Sierra Nevada (Lone Pine), CA	6 feet	15–20 feet	65–100 miles	7.4–7.8 (est.)
1915 Tobin Range (Pleasant Valley), NV	18 feet	less than 3 feet	37 miles	7.1 (est.)
1932 Cedar Mountain, NV	6 feet	less than 6 feet	39 miles	7.2 (est.)
1954 Fairview Peak and Stillwater Range (Dixie Valley), NV	12 feet	12 feet	60 miles	6.8–7.1
1959 Hebgen Lake, MT	20 feet	6 feet	17 miles	7.3
1983 Borah Peak, ID	9 feet	negligible	21 miles	6.9
1992 Landers, CA	6 feet	18 feet	44 miles	7.3
1999 Hector Mine, CA	negligible	17 feet	22 miles	7.1

[1] Data are from the U.S. Geological Survey Earthquake Hazards Program: http://earthquake.usgs.gov/regional/states/historical.php. Note that the table only includes quakes of Richter magnitude 6.8 or larger. Quakes of smaller magnitude have occurred more frequently.

[2] Richter magnitude is based on the size of seismic waves recorded at a seismograph. The scale is logarithmic, so that an increase of 1.0 on the scale represents ten times more ground shaking. (For instance, a magnitude 7.3 quake causes 10 times more shaking than a 6.3.) Typically, an earthquake needs to have a Richter magnitude of 3.0 or greater for people to feel it, and 5.0 or greater to cause damage. Magnitudes greater than 7.0 can cause considerable damage even miles from the epicenter. There is no upper limit to the Richter scale, but because there is a limit to the strength of rock (meaning the amount of strain energy that rock can store before it breaks), magnitudes greater than 9.0 are very rare. Modern seismographic networks have been in place only since about 1950. Richter magnitudes for quakes older than 1950 are estimated from displacement, scarp length, and eyewitness accounts.

Reno–Lake Tahoe area is clearly at risk from the Genoa fault. But a greater potential disaster may be waiting at Salt Lake City, in the Intermountain Seismic Belt on the opposite side of the Great Basin.

Arguably, no urban area in the United States has a more stunning backdrop than Salt Lake City. On jet approach to the airport, everyone cranes for the window to watch the ramparts of the Wasatch Range pass by, below the plane at first and then above it as the jet settles toward the runway. The Wasatch Range and the Sierra Nevada face one another from the opposite sides of the Great Basin, and although the Wasatch is smaller, its fault-bounded escarpment is every bit as spectacular, especially where it soars more than a mile above Salt Lake City.

The Wasatch Range exists because of the Wasatch fault, which divides the thin, stretched, earthquake-prone crust of the Basin and Range to the west from the thicker, more stable crust of the Rocky Mountains and Colorado Plateau to the east. In many places, the Wasatch fault is marked by a scarp so sharp and clear that it looks as if someone had chalked the seam between basin and range. (Figure 7.7 shows a close-up of the

TABLE 7.1B   Prehistoric Earthquakes on the Genoa Fault (Carson Range) and the Wasatch Fault
(Wasatch Range)

Prehistoric Quakes[1]	Vertical displacement	Horizontal displacement	Scarp length	Richter magnitude[2]
Genoa fault, Carson Range, NV: two prehistoric quakes, ~2100 years ago and ~550 years ago	10–18 feet	negligible	15–45 miles	7.2–7.3 (est.)
Wasatch fault, Wasatch Range, UT: 20 prehistoric quakes on five fault segments, oldest ~6000 years ago, youngest ~500 years ago	6–20 feet	negligible	20–40 miles per segment	7.0–7.5 (est.)

[1] Prehistoric quake data are from Ramelli and others 1999 for the Genoa fault, and from McCalpin and Nishenko 1996 for the Wasatch fault, supplemented with more recent data from the Utah Geological Survey website: http://geology.utah.gov/utahgeo/hazards/eqfault/index.htm.

[2] For an explanation of Richter magnitude, see Table 7.1A.

scarp near downtown Salt Lake City.) You need only compare figure 7.5 to the bottom photo in figure 7.6 to see that the Wasatch Range classifies as a fast-rising mountain. Yet the Wasatch fault hasn't ripped loose with a major quake since at least 1847, when Mormon pioneers settled the Salt Lake Valley. Since then, the population of the greater Salt Lake City metropolitan area has skyrocketed, and today more than 1.6 million people—80 percent of Utah's population—live strung out along the base of the Wasatch Range, many of them within a mile or two of the Wasatch fault. Some of the most expensive homes in Utah lie a stone's throw from the fault. Prime real estate it is too, with the Wasatch Range soaring up from backyards and panoramic views of the Salt Lake Valley spreading out from front windows. Enjoy it while it lasts, folks.

To evaluate the threat of the Wasatch fault, Utah geologists have dug more than seventy trenches across the fault in search of evidence for past earthquakes. Two questions have driven their quest: how often do large earthquakes happen along the fault, and how large can those quakes be? They've found answers where the fault has split old soil beds or layers of plant debris. By age dating those layers using radioactive carbon, and by measuring the amount that the fault has shifted those layers, the geologists have pieced together a history of earthquakes along the Wasatch fault that reaches back more than 6,000 years (figure 7.8).

The Wasatch fault, it turns out, is not continuous; rather, it is divided into about ten end-to-end segments, each about twenty-five miles long and each moving somewhat independently of the other segments. It so happens, unfortunately, that the five most earthquake-prone segments (shown on figure 7.8) all border the greater Salt Lake City metropolitan area. At some level, this should be no surprise; the activity of those five

FIGURE 7.7

The scarp of the Wasatch fault about two miles north of downtown Salt Lake City near Warm Springs. The exposed fault plane forms a cliff nearly forty feet high, and its surface is etched with slickenlines: parallel grooves scraped on the fault plane as the Wasatch Range lurched upward. This site lies at the north end of the Salt Lake City segment of the Wasatch fault shown in figure 7.8.

fault segments is precisely why the Wasatch Range is so steep and spectacular near Salt Lake City.

Within the last 6,000 years, those five fault segments have together fired off at least twenty earthquakes of estimated Richter magnitude 7.0 or greater. That works out to an average of 300 years between large earthquakes. This doesn't mean that a large quake happens predictably every 300 years, of course; far from it. As the time line in figure 7.8 makes clear, the timing of these large quakes has been irregular, with great variation in the time gaps between quakes. Because it has now been about 500 years since the last major Wasatch fault earthquake, some popular press reports have declared that the fault is "overdue." That isn't the right way to think about it, because it implies that large Wasatch quakes have a consistent recurrence interval. Figure 7.8 shows that over the last 6,000 years there have been six gaps of 500 years or longer between large quakes. Still, the data point to one inescapable conclusion: another massive earthquake will whipsaw the Wasatch Front one day, and with potentially calamitous results.

· · ·

July 24, 1847: He comes down through the forests of the Wasatch Range, followed by his minions hauling their wagons along the boulder-clogged riverbed. He is the leader

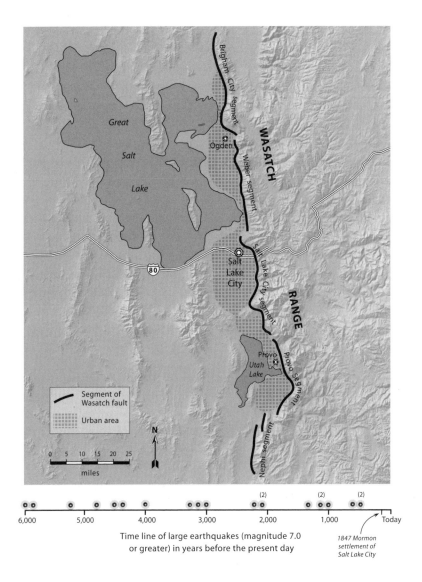

FIGURE 7.8

The 240-mile-long Wasatch fault consists of about ten end-to-end north–south segments. Shown here (top) are the five most active segments, all of which border the most densely urbanized areas of the Wasatch Front. Geologists have unearthed evidence for at least twenty earthquakes of estimated Richter magnitude 7.0 or larger along these five segments during the past 6,000 years. The time line (bottom) shows the dates of these twenty quakes, demonstrating the inevitability of more large quakes in the future. Salt Lake City today joins Los Angeles and San Francisco in the U.S. Geological Survey's top tier of national urban areas at highest risk of a catastrophic earthquake.

of an exodus, and his purpose is simple: to put as much space as possible between his people and the rest of the United States. Separation by vast and rugged geography will, he hopes, let his followers pursue their peculiar brand of Christianity in peace. He turns the last curve in the canyon and sees the well-watered valley sloping down to the shores of Great Salt Lake. "This is the place," he declares—in Mormon legend at least.[4]

The Mormon leader Brigham Young had had his eye on the Salt Lake Valley as a possible settlement site for his Latter-Day Saints since 1844, when he had read John C. Frémont's glowing accounts of the fertile valley watered by streams tumbling down from the Wasatch Range. It was the Wasatch Range, Young knew, that made the Salt Lake Valley such an ideal settlement site. In the otherwise arid and forbidding Great Basin, the Wasatch Range traps water as winter snow and doles it out reliably all year long. Yet the water-giving Wasatch exists because of the ongoing geologic violence that built the Basin and Range and the Great Basin within it. Salt Lake City's eventual geologic demolition is a foreseeable consequence of settlement in a region wracked by the growing pains of mountains.

NOTES

1. The U.S. Geological Survey brass benchmark on Mount Whitney puts the elevation at 14,494 feet, and that's the value that you'll see on many maps. The current official elevation of 14,505 feet, established recently by the U.S. National Geodetic Survey, is based on a more precise measurement of the geoid: an imaginary baseline represented by extending a smooth sea-level surface continuously underneath the continents. The geoid represents the baseline against which continental elevations are measured. The Geodetic Survey's recent adjustment to the geoid baseline has added several feet to many western U.S. peaks.

2. The Lone Pine fault is part of the Eastern Sierra fault system, which, along with the Walker Lane fault system, divides the Sierran Plate from the North American Plate, as I explained in chapter 5. Whenever one of the faults in the Eastern Sierra system or the Walker Lane system shifts, it lets the Sierran Plate slide northwest relative to the land to the east, as shown in figure 5.6.

3. Throughout this chapter, whenever I speak of mountain ranges being uplifted during earthquakes, what I mean is that the elevation difference between a range and its neighboring basin has increased. Whether that increase was produced by the range *rising* versus the basin *dropping*, as measured against some external reference point, is often hard to determine. Technically, it's probably more correct to think of the basins in the Basin and Range Province dropping rather than the mountains rising, since the overall evolution of the region has been one of collapse since the days of the high Nevadaplano (see figure 6.6). But I will continue to write of the mountains rising since that is more visually intuitive.

4. Although it is one of the most memorable phrases in western history, Brigham Young's four-word declaration—supposedly uttered when he first saw the future site of Salt Lake City on July 24, 1847—is almost certainly a myth, gaining canonicity through repetition and its

appeal to Mormonism's sense of predestination. Contemporary diaries make no mention of Young's words, nor did Young himself ever claim them. The phrase first appeared in 1880, three years after Young died, when Wilford Woodruff recalled his experience of being with Young on that day thirty-three years earlier when the Mormon pioneers first entered the Salt Lake Valley. Today, This Is the Place Heritage Park, at the mouth of Emigration Canyon near downtown Salt Lake City, commemorates the spot where Young didn't say his famous words.

8

WEALTH AND MAGMA

The Great Basin—dry, rugged, earthquake-rattled, and mostly empty of people—is nonetheless a land of staggering geologic wealth. California, the Golden State, is misnamed. Nevada, the Silver State, today produces far more gold annually than California ever did, even during the gold rush. Nevada today accounts for three-fourths of all U.S. gold production and nearly 10 percent of world gold production. And for every ounce of gold, Nevada generates nearly two ounces of silver, thirty ounces of copper, and quantities of lead, zinc, iron, molybdenum, mercury, uranium, and tungsten. You can't go anywhere in Nevada without running into evidence of mining: angular head frames, gaping holes of adits (abandoned tunnel entrances), deltaic piles of varicolored waste rock on the mountainsides. Today, as for more than a century, boomtowns rise where new ore is found and ghost towns sag where the ore has run out. Ore, by definition, is rock that holds a profitable concentration of valuable metals. Nevada may hold more ore than any other U.S. state.

One irony of the California gold rush is that the pioneers of '49, as they hurried west across Nevada, passed by far more gold—and other valuable metals—than they would ever find in California. But that soon changed. In 1859, thousands of out-of-work California miners headed east to Nevada to mine silver and gold at the newly discovered Comstock Lode in the Virginia Mountains a few miles south of present-day Reno.

The Comstock Lode launched Nevada's mining boom. The tiny town of Virginia City, astride the Comstock, mushroomed into a burg of 30,000 people, 110 saloons, twenty theaters, and uncounted brothels and opium dens. In 1862, a twenty-six-year-old failed prospector took a job at the local Virginia City newspaper, the *Territorial Enterprise*, where he began to write articles about the Comstock Lode under a soon-to-be-famous pen name. "Vice flourished luxuriantly," Mark Twain remembered of Virginia City. "The saloons were overburdened, so were the police courts, the gambling dens, the brothels and the jails—unfailing signs of high prosperity in a mining region—in any region for that matter. Is it not so? A crowded police court docket is the surest of all signs that trade is brisk and money plenty."[1]

The source of Virginia City's wealth ran right under Main Street. "All along under the center of Virginia and [the neighboring town of] Gold Hill, for a couple of miles, ran the great Comstock silver lode—a vein of ore from fifty to eighty feet thick between its solid walls of rock," Twain wrote. At any given moment, close to half of the city's population was underground among the drifts and tunnels, and Twain recalled how "often we felt our chairs jar, and heard the faint boom of a blast down in the bowels of the earth." From 1859 to 1878, some $400 million in silver and gold poured out of the Comstock mines (roughly $7.2 billion in today's dollars[2]). It was a heady time. "Joy sat on every countenance," Twain remembered, "and there was a glad, almost fierce, intensity in every eye that told of the money-getting schemes that were seething in every brain and the high hope that held sway in every heart." By 1881, though, the boom had busted. Virginia City today survives by mining tourist dollars.

Approaching Virginia City from the south, you wind steeply uphill into the Virginia Mountains, passing abandoned tunnels, tailings, and head frames. The road crests 6,300 feet above the sea at Twain's adopted city, "roosting royally midway up the steep side of Mount Davidson." Immediately, you're funneled onto the once historic but now merely histrionic Main Street, a blaring strip of tourist-trap shops offering, among the kitsch, an astonishing variety of Mark Twain statuettes. Push past this facade, though, and you'll find that much of the town retains its singular history. Perched on the high western edge of town is Millionaire's Row, a string of opulent nineteenth-century homes built with Comstock wealth. Nearby sits the original Virginia City & Gold Hill Water Company building—command center for one of the greatest unsung engineering feats in U.S. history. During the 1860s, the growing mines and population of Virginia City quickly outstripped available water supplies. The Sierra Nevada, ten miles to the west, held abundant water, but getting the water down from there, across the low Washoe Valley, and then 1,300 feet uphill to Virginia City was a daunting problem. The solution came in laying a riveted steel pipe across the valley in the form of a large, upright U, with the west end in the Sierra Nevada and the east end at Virginia City. Water was channeled into the west end of the pipe and plunged nearly 2,000 feet to the valley floor, where the immense pressures were contained by triple-riveting each joined section of pipe. The east end of the pipe, near Virginia City, lay 465 feet lower than the west end. Backed by the pressure of the higher west end, the water gushed out of the east end, where flumes sent it to boilers that generated steam for running pumps and pulling ore carts in the bustling Comstock mines.

The rock around Virginia City testifies to the geologic forces that made the Comstock Lode, and it serves well—in broad outline at least—as a model for many Great Basin ores.

Color is the most evident change that happens when rock becomes metal-rich. Looking down on Virginia City from the air, one sees that it sits on a strip of yellow-orange rock about a half-mile wide and four miles long that runs north–south along the eastern slope of Mount Davidson. Within this yellow-orange strip lies the gutted Comstock Lode—which, despite the name, was not a single lode (that is, one vein of metal-rich rock) but rather an eighty-foot-wide swath of many veins that trended roughly north to south beneath Virginia City and Gold Hill. The rock within that yellow-orange strip is andesite: the remains of lava vented from long-vanished volcanoes. It's the same rock that makes up the surrounding mountains, but within the strip metal-enrichment has changed it in several ways. First, the color has shifted from an original dark brown to a mottled yellow-orange. Second, the rock is thoroughly busted up with fractures and so crumbly that you could dig it up with a shovel—as many Comstock miners did.[3] Third, the andesite is shot through with veins of quartz (figure 8.1).

Each of these clues reveals a piece of the Comstock's metal-enrichment story. First, the busted-up rock tells us that the Comstock Lode lies within a north–south zone of faults, whose grinding movements have shattered the rock. Second, the color change reflects chemical reactions set off when hot, underground waters flowed through the shattered rock, dissolving existing minerals while depositing new ones. Third, the abundant quartz veins indicate that these groundwaters carried dissolved silica—the most abundant chemical compound in the Earth's crust. This is important, because dissolved metals often come along with silica in hot underground waters. When silica precipitates from hydrothermal waters to make quartz veins, the metals often drop out of solution too. You'll rarely *see* the metals in any Great Basin ore; they're typically microscopic or locked up within minerals that don't look much like the metals in their pure state. For instance, most silver in the Comstock Lode occurs as acanthite (silver sulphide)—a soft, dark-gray mineral that looks nothing like native silver. Before 1859, gold miners in the Virginia Mountains cursed this ugly acanthite. When wet, it turned to a sticky sludge that plugged up their flumes, interfering with their search for gold. Fortunes in acanthite were shoveled aside until two brothers named Groch discovered the hidden silver within, unleashing the silver rush to the Comstock Lode in 1859.

.   .   .

Nearly all Great Basin ores, including the Comstock Lode, owe their existence to groundwater. Circulating groundwater acts like a chemical broom, sweeping up atoms of gold,

FIGURE 8.1

In this outcrop near Virginia City (top), I'm standing at the contact between the unaltered andesite that makes up most of the Virginia Mountains and the hydrothermally altered andesite that contains the Comstock Lode. The altered andesite (bottom) is crumbly and riddled with fractures caused by movements on nearby faults. The fractures allowed hydrothermal waters to percolate through the rock, changing its color and filling the fractures with quartz veins, some of which hold precious metals. Some Comstock quartz veins are more than a foot thick.

silver, copper, and other metals and depositing them where the water percolates through zones of fractured rock. But groundwater alone doesn't explain the Great Basin's mineral wealth. (After all, groundwater occurs everywhere, but ores do not.) Rather, it is the conjunction of groundwater and *magma*.

To dissolve and mobilize metals in quantity, groundwater needs to be very hot—and that requires magma underground. Moreover, magma turns out to be an important source of the metals in many ores. As groundwaters circulate, they pick up metals both from the surrounding rock and from bodies of magma through which they pass. Most magma contains only scant quantities of metal atoms per unit volume of molten rock. But as magma solidifies into rock, any remaining liquid becomes progressively enriched in metal atoms. That's because most metal atoms don't readily form chemical bonds with other atoms in magma. Therefore, as the magma crystallizes, the metal atoms are often left out of the growing mineral grains, like the geeky kids on the playground who aren't picked to join the teams. As the magma converts into rock, the left-out metal atoms accumulate in ever-greater concentrations in the shrinking pool of molten rock. The result is that the very last dregs of magma are often super-enriched in metals. These dregs of concentrated wealth often solidify in the uppermost areas of a cooling pluton, or in the crust right above it, because magma (being less dense than solid rock) rises where it can. In this way, metals often collect into rich concentrations above and around plutons. If the pluton is uplifted, say, with a mountain and erosion strips away most of the rock overhead, then miners can excavate the carapace of metal ore surrounding the pluton. Alternatively, the metals may be picked up from the pluton by scalding groundwaters and whisked off to a site of precipitation, such as a fault zone above the pluton (figure 8.2).

Bottom line: if you want to make metal ore in quantity, you need lots of magma to invade the crust. The Great Basin, it turns out, has experienced two great magmatic invasions, both of which seeded the region with metal ores. The first wave began about 140 million years ago, near the end of Jurassic time, and peaked about 100 to 90 million years ago, during the middle of Cretaceous time. The second wave—shorter but more intense than the first, and responsible for more ore—lasted from about 43 to 21 million years ago, during the middle of Cenozoic time. To understand why magma (and thus ore) came in these two separate waves, we need to revisit North America's history of westward migration.

North America's tectonic journey of the last 200 million years—a journey that has taken it away from Europe and Africa to open the Atlantic Ocean—has caused the continent to override several oceanic plates along its western edge. The wreckage of some of those plates today forms accreted terranes throughout British Columbia, Washington, Oregon, and California (including those great terrane belts in the western Sierra Nevada that we visited in chapter 2). About 140 million years ago, the largest of these old seabed plates—the Farallon Plate—began plunging under the continent's advancing

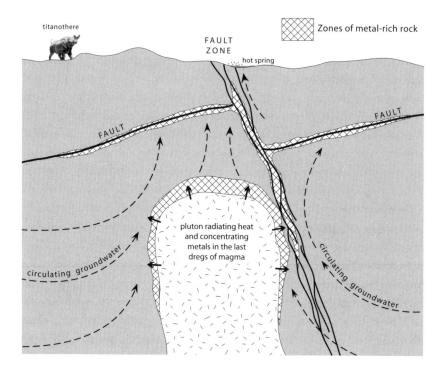

titanothere

FAULT ZONE

hot spring

Zones of metal-rich rock

FAULT

FAULT

pluton radiating heat and concentrating metals in the last dregs of magma

circulating groundwater

circulating groundwater

FIGURE 8.2

Most Great Basin metal ores occur in and around plutons—large blobs of magma that rise up from the mantle and stop deep in the crust. As a pluton solidifies, metal atoms are concentrated in the remaining molten rock, forming metal-rich zones above and around the pluton. Circulating groundwaters heated by the pluton pick up metals from these zones and the surrounding rock and then rise buoyantly, often along fault zones, where they cool and deposit metal-bearing minerals, often with abundant quartz. The titanothere (an extinct rhinoceros-like mammal once common in the western United States) pegs the diagram at mid-Cenozoic time, when the second of two great waves of magmatism and ore formation swept through the crust of the future Great Basin. (The titanothere is not to scale; plutons can be more than a mile in diameter and form several miles belowground.)

edge. Gobs of magma began to well up into the crust where the Sierra Nevada and the Great Basin are today. Where this magma reached the surface, it vomited lava and volcanic ash across the landscape. But of greater significance for Great Basin ore production was the magma that *didn't* reach the surface. Many of the molten blobs stalled deep belowground, somewhat like helium balloons bumping into a ceiling. There they congealed into plutons. Magma production peaked about one hundred million to ninety million years ago, as hundreds of plutons glommed together to form the Sierra Nevada Batholith (chapter 4), while others welled up to the east, into the crust of the future Great Basin. The plutons heated underground waters, which swept up metals and silica from the plutons and the surrounding crust. Pregnant with metals, the waters rose buoyantly along permeable pathways, favoring fault zones above and around the plutons. In these fault zones, the hot waters cooled and dumped their metallic loads as metallic lodes. Thus was seeded the first wave of Great Basin ore—and, to the west, the gold of California's Mother Lode (chapter 2).

By eighty million years ago, the molten upwellings that had seeded the first wave of Great Basin ore had sputtered out. The reason most likely relates to the flat subduction of the Farallon Plate—an idea that I introduced in chapter 6 and which I'll develop and support with evidence in chapter 12. Between eighty million and forty-five million years ago, the Farallon Plate seems to have slid horizontally beneath the continent to push up the Rocky Mountains (figure 8.3 top). Little magma formed during this interval because the plate lay above its magma-generation depth in the mantle.[4] But by forty-three million years ago, the Farallon Plate seems to have restored itself to normal subduction mode. It began cranking out magma again, and on an awesome scale. Like the start of a hot-air balloon race, hundreds of molten blobs took off from the mantle above the Farallon Plate (figure 8.3 middle). Where this magma pooled belowground, it cooked up hydrothermal waters to seed the second wave of Great Basin ore, from about forty-three million to twenty-one million years ago.

Not all of the magma that rose up during this second wave of ore formation stayed underground. Prodigious amounts punched its way up to daylight. There, it unleashed volcanic Armageddon across the future Great Basin. The scope and violence of this mid-Cenozoic volcanic episode—coincident with the second wave of metal ore formation—was so extraordinary that it's worth a closer look.

From forty-three to twenty-one million years ago, wave after wave of volcanic ash and lava blasted forth across the future Great Basin. It came out of calderas—huge volcanic craters as much as fifty miles across, formed where magma erupts with such violence that the land surface collapses into the underground space where the magma used to be. When most of us think of a volcano, we think of a cone-shaped mountain. But caldera eruptions don't make mountains. They empty the molten guts beneath mountains so that the mountains collapse to form gigantic holes (figure 8.4). Most of the Great Basin's calderas did not form singly but rather as overlapping clusters called caldera complexes. Most of these caldera complexes were subsequently busted up as the crust sundered to form the Basin and Range (figure 8.3 bottom), so that parts of the same calderas sometimes occur in different mountain ranges. When we put these Humpty Dumpty calderas back together again and place them on a map, we see a most interesting pattern. The calderas get progressively younger to the southwest. In other words, the caldera eruptions didn't burst forth everywhere at once across the future Great Basin. Instead, the eruptions formed an advancing wave that swept from northeast to southwest. The pattern must reflect the way that the Farallon Plate restored itself to normal subduction after it scraped flat under the continent to hoist the Rocky Mountains. Perhaps the plate swung back down into the mantle like an opening trap door

~ 80 - 45 million years ago: Flat subduction of the Farallon Plate pushes up the Rocky Mountains; erosion of gold from the Nevadaplano sends auriferous gravels west to California.

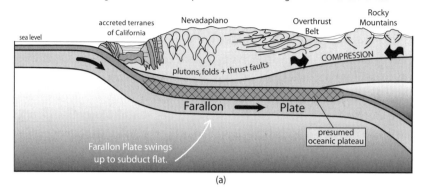

(a)

~ 43 - 21 million years ago: Farallon Plate resumes normal subduction; rising magma seeds the crust of the future Great Basin with metal ores while unleashing cataclysmic caldera eruptions.

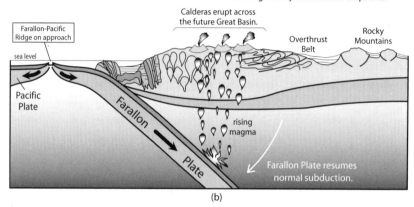

(b)

Last 20 million years: North America overrides Farallon-Pacific Ridge. Crust stretches to form the Basin and Range Province and collapses to form the Great Basin. Metal ores are exposed in the new mountain ranges.

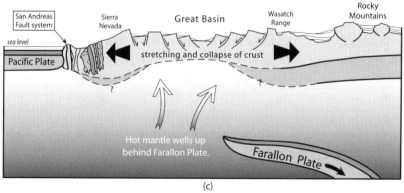

(c)

FIGURE 8.3

How the plate tectonic history of western North America and the Farallon Plate generated metal ores in the Great Basin.

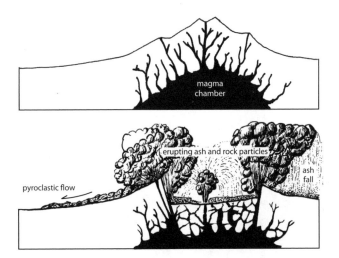

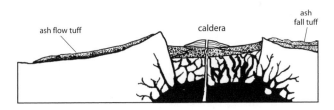

FIGURE 8.4

Volcanic calderas form when volcanoes erupt with such violence that the land collapses into the partially emptied magma chamber below the volcano, forming a crater miles across. Erupted material blasts out in two main ways: as dense, red-hot clouds called pyroclastic flows that rush downhill like scalding avalanches, or as particles of rock and ash (tiny bits of solidified magma; no relation to wood ash) that rise miles into the sky before falling back. When a pyroclastic flow stops and settles, the glowing-hot particles weld themselves together into a hard, dense rock called ash flow tuff. When particles of cool ash settle from the sky, they form a powdery, soft rock called ash fall tuff. Basin and Range faulting has broken up most of the mid-Cenozoic calderas of the Great Basin. Younger and better-preserved examples of calderas in the western United States include Oregon's Crater Lake, the Yellowstone Caldera of Yellowstone National Park, and Long Valley Caldera near Mammoth, California.

hinged somewhere southwest of the Great Basin. Whatever the details, the result was a twenty-two-million-year episode of volcanic hell—complete with fire.

The calderas hurled forth scalding avalanches of red-hot rock particles called pyroclastic flows, and spewed tumescent clouds of ash miles into the sky. Today, these volcanic rocks stack up thousands of feet thick in some Great Basin mountain ranges, forming benign bands of ochre, pink, and gold that catch the Sun's rays with a postcard glow. Travel back in time to this volcanic wasteland, and your eye takes in a landscape

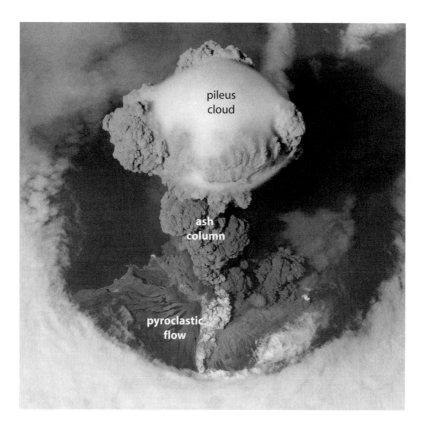

FIGURE 8.5

This eruption may mimic—albeit at a smaller scale—what a Great Basin caldera eruption might have looked like if you had been able to look down from outer space during mid-Cenozoic time. The image, taken by astronauts aboard the International Space Station, captures the explosive eruption of Sarychev Volcano in the Kuril Islands on June 12, 2009. The mushroom-shaped column of erupting ash has punched a wide hole in the surrounding clouds. Air pushed upward ahead of the ash column has cooled and condensed to form a short-lived pileus cloud on top of the ash column. Meanwhile, a pyroclastic flow—a dense avalanche of hot gas and glowing rock particles—roars down the volcano's flanks.

as desolate as Mars. In every direction lie sheets of powdery ash, cloaking the world like some ghoulish Christo project. For years, the land looks dead, except perhaps for a few flies that buzz in from far away to feast on ash-fried meat. Eventually, some hardy wind-borne seeds take root. After several decades, a thin soil has formed. More plants take hold, and animals follow. Then, a rumbling. A drumroll of small earthquakes lasting several weeks signals the buildup of magmatic pressures underground. The land bulges. Cracks appear. Then, with a blast that ruptures every available eardrum, a column of glowing rock blows sky high, expanding upward and outward as a colossal mushroom cloud. (Figure 8.5 shows a modern example.) Inevitably, gravity must reclaim this cloud. Thousands of tons of glowing hot rock particles reach their zenith and

FIGURE 8.6

The deadly power of pyroclastic flows. The top photograph shows the remains of the city of Saint-Pierre on the Caribbean island of Martinique, wiped out in May 1902 by a pyroclastic flow from Mount Pelée. The bottom photograph shows a plaster cast of a pet dog, still wearing its collar, suffocated by one of the pyroclastic flows that destroyed Pompeii in A.D. 79. Hot volcanic ash settled and hardened around the Pompeii corpses, which then rotted, leaving hollows in the rock that archeologists eventually discovered and filled with plaster to make casts.

begin to fall back. The collapsing cloud hits the Earth's surface at more than a hundred miles per hour and angles outward along a roiling incandescent front—a pyroclastic flow (pyro = fiery; clastic = broken pieces). Meanwhile, other pyroclastic flows have burst sideways out of the flanks of the caldera, as if shot from a cannon. Death is mercifully quick—faster than for a boiled lobster, for the temperatures are eight times hotter. Mammals turn to charred steak, birds to cinders, trees to embers. Then, over the following months and years, life tries once again to repossess the land.

Pyroclastic flows have incinerated thousands of people through human history. Famous examples include 28,000 killed on the Caribbean island of Martinique in May 1902, and the A.D. 79 eruption of Mt. Vesuvius, immortalized by people and animals preserved in poses of gruesome death in Herculaneum and Pompeii (figure 8.6). Yet these historic pyroclastic flows were puny compared to the prehistoric ones that once scalded the Great Basin. I have in my rock collection a piece of ash flow tuff (the rock that forms when a pyroclastic flow solidifies) from the twenty-three-million-year-old Bates Mountain Tuff of central Nevada. It contains fragments of pumice that were originally more-or-less spherical but which were so hot—and thus soft—when the pyroclastic flow settled that they flatted into pancakes under the weight of the overlying rock. Imagine being engulfed by a volcanic cloud so hot that the pieces of rock within it are as soft as Play-Doh. The Bates Mountain Tuff contains roughly 250 cubic miles of volcanic ash—enough, if spread evenly, to cover every square mile of Nevada twelve feet deep. By comparison, the 1883 eruption of the Indonesian island of Krakatoa—a blast so violent that it decapitated a five-mile-wide island and killed 36,000 people (mostly from tsunami waves)—involved a measly 4.3 cubic miles of lava and ash. Multiply Krakatoa *sixty times,* and you have the Bates Mountain Tuff—which, in turn, represents just one of dozens of volcanic units spewed forth during the mid-Cenozoic caldera eruptions of the Great Basin.

While all of this volcanic mayhem was bursting forth aboveground, much of the magma that fueled it stayed belowground, forming plutons that seeded the second wave of Great Basin ore. Given the volume of magma involved, you won't be surprised to learn that more ore formed during this mid-Cenozoic wave than during the earlier Jurassic-Cretaceous wave. The second wave of ore formation created the Carlin Trend, Nevada's most productive gold belt (named for an elongate belt of gold mines in the region around Carlin, Nevada). The second wave also planted the ore at Bingham Canyon, Utah (site of the world's largest open-pit copper mine), and deposited the silver and gold in western Nevada's Comstock Lode. If you live in the United States, the odds are good that at least some of the gold in your rings, the silver in your drawers, and the copper in your pipes came from a Great Basin mine. And if it did, the odds are very good that that ore formed during mid-Cenozoic time, collecting in rich veins underground while pyroclastic flows charbroiled the land above and volcanic ash settled from the sky like toxic snow.

NOTES

1. Samuel Clemens went west in 1861 after losing his job as a riverboat pilot when the Union Navy blockaded the Mississippi River at the start of the Civil War. After a year of fruitless prospecting in western Nevada, he began work at the *Territorial Enterprise*. It was there, on February 23, 1863, that he began signing articles as Mark Twain. His rise to national fame began in 1865, after he moved to California and published *The Celebrated Jumping Frog of Calaveras County*.

2. Based on historical currency conversion tables by Robert C. Sahr, Political Science Department, Oregon State University, Corvallis.

3. Deadly cave-ins were common in early Comstock mines because of this crumbly ore rock. In 1860, an engineering innovation called square-set timbering put a stop to most cave-ins. Stout lumber was brought over from the Sierra Nevada and fastened together into open cubes about six feet on a side. As the ore was removed, the wooden cubes were inserted into the voids. Mark Twain described the result on one of his underground visits. "Over their [the miners'] heads towered a vast web of interlocking timbers that held the walls of the gutted Comstock apart. These timbers were as large as a man's body, and the framework stretched upward so far that no eye could pierce to its top through the closing gloom. It was like peering up through the clean-picked ribs and bones of some colossal skeleton."

4. A subducting plate won't generate magma until it reaches depths of about sixty to eighty vertical miles in the mantle. Flip back to chapter 4 for an explanation of how and why magma forms during subduction.

9

WATER AND SALT

Along the south bank of the Snake River in southern Idaho lies a black boulder of basalt twelve feet across. It sits in a grassy hollow several hundred yards uphill from where the river glides by in its canyon. The ground around the boulder was a favorite resting spot for pioneers on the Oregon and California trails. While their animals grazed along a nearby creek, some pioneers chiseled their names into the surface of the boulder. Weather and vandalism have obscured many of the names (a fence and roof protect the boulder today), but dozens of others are still visible. *E Wilson 1859; H Chestnut Aug 20 1862; T J Wilcox July 29 1872 from Iowa; J M Hepler July 1882,* and so on. It is called Register Rock.

I don't know whether any of these pioneers pondered how such a large boulder came to rest at this spot, but anyone who thinks about it for a moment will recognize the problem. Some math based on the boulder's dimensions and the density of basalt shows that Register Rock weighs at least fifty tons. Yet it lies far from the Snake River, and far from any precipice from which it could have tumbled. It's not a meteorite, so it didn't fall from the sky. Curiously, the boulder has no sharp corners. Instead, the corners are smooth and round like on a beach pebble. This can only mean that the boulder *rolled* some distance to get here. Yet what force of nature could have rolled a rock three times heavier than a Greyhound bus?

Register Rock, it turns out, is not alone. At Massacre Rocks, two miles upstream from Register Rock, basalt boulders bigger than SUVs lie strewn across the bottom-

FIGURE 9.1

An example of the so-called melon gravels along the Snake River Valley in southern Idaho. When the Bonneville Flood swept through here some 14,500 years ago, it rolled these immense basalt boulders and smoothed their corners and edges, so that they look like large versions of pebbles tumbled by a mountain stream.

lands. These, too, have had their corners smoothed off. Downstream along the Snake River, you'll find gravel bars more than a hundred feet high and several miles long packed with watermelon-shaped boulders up to eight feet across (figure 9.1). These so-called melon gravels got their name when jokesters put up signs on the nearby roads advertising petrified watermelons free for the picking. *Take one home to your mother-in-law!* one sign suggests.

It was a flood that tumbled these boulders—but not a normal one. You won't find a river anywhere on Earth today, even in flood stage, that can roll rocks that big. This was a swollen freak of a flood, an abnormality of water that once roared 500 feet deep down the Snake River Valley to the Columbia River and on to the Pacific. It may have been the biggest freshwater flood in the history of the world as measured by the total volume of water released. Usually, great floods go with great rainstorms. But no storm caused this flood. Rather, the water came from Lake Bonneville, an immense lake that, as recently as 15,000 years ago, stood close to one-fifth of a mile deep over Salt Lake City. After disgorging 1,150 cubic miles of water to create the Bonneville Flood, Lake Bonneville gradually evaporated in the warming post–ice age climate. What's left is a big, salty puddle called Great Salt Lake.

13,000 B.C.: He comes down through the snow-dusted forests of the Wasatch Range. You would smell him, or hear him, long before you saw him. Normally, it wouldn't be wise to announce one's presence so boldly here, given what else stalks these woods. But keeping a low profile isn't an option for him, and he doesn't need one anyway. Round footprints as big as dinner plates dot the snow behind him, and branches fifteen feet above the ground bend and snap back as he passes by. He turns the last curve in the canyon and sees a sprawl of blue as large as Lake Michigan, dotted with wooded islands that cut sharp profiles into the sky. He crosses the delta, formed where the icy river has slewed sand and gravel into the lake, and walks into the water. Breakers dash cold against his wooly legs and against the steep mountain front to his left and right. There, on the mountainside, the waves are cutting a shoreline notch (one that stands out clearly today on the face of the Wasatch Range high above downtown Salt Lake City). He drops his agile trunk into the water and curls it up to drink.

He is not alone. Other charismatic megafauna drink from the shores of Lake Bonneville. There are saber-toothed cats, whose scimitar teeth can quickly slice open trachea and jugular veins. There are packs of dire wolves. There are short-faced bears, half again as large as the largest living grizzly, the most powerful predators to have stalked the continent since the days of *Tyrannosaurus rex*. (Short-faced bears may be the largest land mammal predators to have ever lived, nearly six feet tall on all fours, twice that standing, and weighing up to 2,200 pounds.) Keeping an eye on these predators are assorted herbivores, including bighorn sheep and bison. There are musk oxen, whose descendants will, in a few millennia, retreat to the Arctic as the continent warms out of the ice age. There are camels and horses, too, both native to North America. They will eventually go extinct on this continent, but not before expanding into Asia across the Bering Land Bridge, exposed by low ice-age sea levels. (In 1540, horses will return to the American West with Francisco Vásquez de Coronado's expedition, and escapees will repopulate their ancestors' homelands.) There are giant ground sloths, sixteen feet tall, whose dung is the largest to have landed in North America since the days of dinosaurs. In fifteen millennia, curious bipeds will dissect these gargantuan coprolites in search of clues about diet, habitat, and ice-age climates. (Ground sloths frequently pooped in dry caves, so their dung, which contains seeds and other clues about local vegetation, is often well preserved.)

Along the lake's northern shore, a deep bay pokes north into what is now southern Idaho. There, the lake spills over into the Portneuf River, which drains north to the Snake River. For several centuries, the lake holds steady at the level of this outlet, near what is now Red Rock Pass in southern Idaho. During this time, waves cut notches around the lake's edges and islands, and pile up immense gravel bars and shingle beaches. Eventually, the outflow at Red Rock Pass cuts down through a layer of tough rock that had been holding back the lake. Beneath it, the outflow finds loose gravel. The gravel begins to give way. The lake begins to pour through the breach. Within a few days, the growing breach is discharging a volume greater than the Columbia River. Soon the flow peaks at a volume five times greater than the Amazon—or nearly the total discharge of all of the rivers in the world today, combined.

The canyon of the Snake River, downstream, is a substantial cleft, as much as a mile wide and 500 feet deep. The flood quickly fills the canyon and brims out onto the uplands in a furious, leaping slurry. Wooly mammoths and other resident beasts are swept away. Car-sized blocks of basalt emit dull booms as they bounce along in the flow. The water whips around bends in the Snake River Canyon like an Olympic luge runner—and nearly as fast, too—vacuuming up millions of tons of bedrock and scooping great alcoves out of the canyon walls. Where the flood pinches through narrows, it accelerates. Downstream of the narrows, where the flow slows, millions of flood-rounded boulders—today's melon gravels—drop out in elongate windrows more than a hundred feet high and several miles long. Immense whirlpools gyrate like aquatic tornadoes in the flow, drilling deep holes into the canyon floor that the Snake River pours over in waterfalls today. The deluge carves immense cataracts on the upland surface above the Snake River Canyon. Today, these fossil waterfalls lie high and dry above the Snake River (figure 9.2).

The flood surges west and collides with the Columbia River. Overwhelmed, the Columbia reverses and flows upstream. (The water doesn't flow uphill, mind you; every direction from the flood is downhill.) Pouring west along the valley of the Columbia, the flood backs up behind the Wallowa Gap in a swirl of foam, splintered logs, and sodden carcasses. The water jets through the Wallowa Gap in a filthy cascade, brims high as it squeezes through the Columbia Gorge, and then pours its turbid mess into the Pacific, staining the ocean more than one hundred miles out to sea.

For several months, the flood roars unabated as Lake Bonneville drains through Red Rock Pass. Eventually, the outflow encounters a layer of resistant rock beneath the loose gravel that had given way. The flow tapers off, and the lake stabilizes at a new level, 350 feet below its pre-flood mark. In all, the Bonneville Flood has released some 1,150 cubic miles of water—enough, if you could spread it out evenly, to cover all of the lower forty-eight U.S. states two feet deep.

If there has ever been a freshwater flood that released more water than the Bonneville Flood, its evidence has so far eluded us. (The largest known flood on Earth was a saltwater flood that took place about 5.3 million years ago, when the then nearly dry Mediterranean Sea refilled as the Atlantic Ocean poured over the Strait of Gibraltar.) When we speak of freshwater floods, there are two standard measures: volume (the total amount of water released during the flood) and peak discharge (the maximum amount of water released at a given moment). The Bonneville Flood's volume of 1,150 cubic miles beats any known freshwater flood. However, its peak discharge (estimated at twenty-one cubic miles per day), falls well short of two other prehistoric floods: the

Interstate 84

cataracts carved
by the flood

FLOW OF
BONNEVILLE
FLOOD

Snake River Canyon

FIGURE 9.2

The marks of the Bonneville Flood along the Snake River. The view looks north, from the air, across the Snake River Canyon near Twin Falls, Idaho. The flood overtopped the canyon (which is more than 300 feet deep here) and spread several miles wide across the uplands. Roiling west at perhaps sixty miles per hour, the flood carved large cataracts on the volcanic basalt of the upland surface. Today, these fossil waterfalls (the largest of which is 800 feet across) stand high and dry above the Snake River.

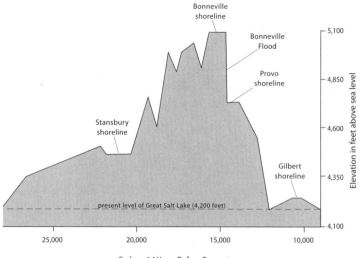

FIGURE 9.3

The rise and fall of Lake Bonneville from about 30,000 years ago to about 10,000 years ago, showing the timing of the four major shorelines (the Stansbury, Bonneville, Provo, and Gilbert shorelines), and the Bonneville Flood.

Missoula Floods of the Pacific Northwest and the Altai Floods of Russia. Both the Missoula and the Altai floods involved the failure of ice dams that had impounded large glacial lakes during Late Pleistocene time. These ice dams formed when glaciers oozed across mountain valleys to block the flow of rivers. Each time an ice dam formed and broke, it released a flood from the lake behind the dam. Several of the Missoula and Altai floods had peak discharges much larger than the Bonneville Flood, but none came close to its volume.

You can read the story of Lake Bonneville and the Bonneville Flood throughout the mountains of northern Utah. Scribed on nearly every mountainside are sharp, horizontal bands, visible from miles away. Climb up to one of these ancient shorelines, and you'll find either a wave-chopped notch or an old gravel beach made of wave-rounded pebbles cemented together like concrete aggregate. The cement is tufa—a pale, crusty form of limestone that frequently precipitates in lakes. Each shoreline records a time when Lake Bonneville's level held steady for a time—probably several centuries or more—so that the waves had time to make their mark. By radiocarbon dating organic materials such as snail shells within these old shorelines, geologists have pieced together the rise and fall of Lake Bonneville (figure 9.3).

During the Pleistocene Epoch, wetter, cooler ice-age climates delivered more water to the Great Basin than left through evaporation. Lakes thus grew in many of today's now-dry basins and sometimes overtopped the basin rims. Lake Bonneville, the largest of these Great Basin lakes, began to fill the Bonneville basin about 30,000 years ago.[1] About 23,000 years ago, the growing lake stabilized for a time, and waves cut the Stansbury shoreline—the oldest of four prominent Lake Bonneville shorelines. The lake then rose again, with occasional reverses, until about 16,000 years ago, when it began to brim over near Red Rock Pass. Now at its maximum possible size, the lake stood more than 1,000 feet deep, with acreage roughly equal to Lake Michigan. For more than a millennium, the lake held steady at this level, while waves cut the highest shoreline, called the Bonneville shoreline (figures 9.3 and 9.4). Then, about 14,500 years ago, the outlet at Red Rock Pass gave way, unleashing the Bonneville Flood. The lake dropped 350 feet before stabilizing again to cut the Provo shoreline.

By the time of the Provo shoreline, the continental ice sheets—those great masses of ice that had oozed south like pancake batter across Canada to bring cool, wet climates to the Great Basin—were in retreat. As the Great Basin warmed and precipitation tapered off, evaporation began to take more water from Lake Bonneville than came in from rivers and rain. The lake began to die of thirst. With one exception, when the lake stabilized for a time to cut the Gilbert shoreline (figure 9.3), the post-flood history of Lake Bonneville has been one of contraction.

As Bonneville dried up, the Earth's crust slowly bobbed upward, adjusting to the removal of the lake's weight. It bobbed up the most where it had sagged the most, right in the center of the lake basin. (Picture the sag you make in a waterbed rebounding when you get up.) Today, the ancient shorelines of Lake Bonneville dome gently upward from the edges of the basin toward the center. The highest shoreline, for instance, rises from 5,100 feet elevation along the margins of the former lake to more than 5,300 feet on mountains that were once islands in the middle of the lake.

·   ·   ·

Great Salt Lake is the runt child of Lake Bonneville. With an average depth of just twelve feet, Great Salt Lake is little more than a salty puddle filling the lowest part of the Bonneville basin. The gradient at the lakeshore is so slight that you may need to walk a quarter-mile or more before your knees get wet. It's a good walk, though; the sand feels smooth on your feet. That's because it's ooid sand—the skin-friendliest sand on Earth. Ooids (rhymes with fluids) are tiny, smooth spheres of white calcium carbonate that precipitate on the wave-agitated lake bed. Rolled between your fingers, they feel like tiny ball bearings.

It might seem strange that a lake fed by fresh snowmelt from the Wasatch Range could be salty, but all rivers carry a fraction of salt dissolved from the rock over which they flow. If a lake overflows at an outlet, the salt passes on through and the lake stays fresh. But once Lake Bonneville fell below the level of its post-flood outlet about 13,500 years ago, it became a so-called terminal lake—one with no outlet. Salt washed into a terminal lake has no way out. When the water evaporates, the salt stays behind. In this

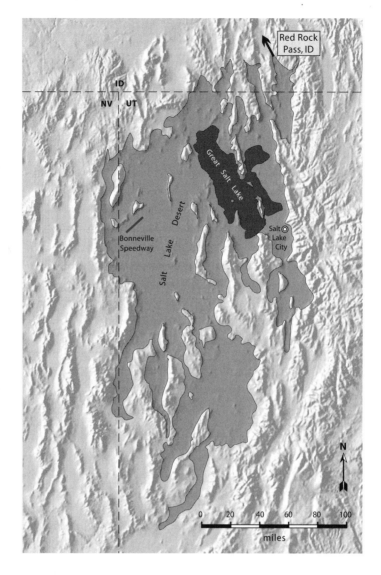

FIGURE 9.4

Map of northwestern Utah and parts of adjacent Idaho and Nevada, showing Lake Bonneville at its maximum size about 15,000 years ago. Note the lake's outlet at Red Rock Pass and the location of today's Great Salt Lake, Salt Lake City, Salt Lake Desert, and Bonneville Speedway.

manner, "fresh" rivers can make lakes that are many times saltier than the ocean. The Dead Sea and Great Salt Lake are classic examples.[2]

Even a lean geek like me, who sinks in a freshwater pond, can bob with fingers, toes, and head exposed in Great Salt Lake, whose waters can be as much as 20 percent denser than fresh water. A skipper at one of the lakeshore marinas once complained to me that his sailboat was hard to steer because it floated so high in the lake. I thought he was pulling my leg, but a walk through the marina showed that the pre-painted factory water lines on the boats stood well above lake level. Since the 1950s, a rock-fill railroad causeway has divided Great Salt Lake into northern and southern arms with distinctly different salinities. The southern arm receives most of the runoff from the Wasatch Range and thus maintains a lower salinity of 12 to 14 percent salt. The water in the northern arm, with almost no river runoff, is 24 to 28 percent salt. (For comparison, the ocean averages 3.5 percent salt.) Only the Dead Sea is more saline (30 to 35 percent salt), which is why it is "dead," at least in terms of animal life. (Bacteria and algae grow abundantly there, since no animals exist to eat them.)

Unlike the Dead Sea, Great Salt Lake is very much alive. The lake has no fish, but more often than not the water swarms with millions of tiny, pink, pulsating bodies of *Artemia franciscana*, the Great Salt Lake brine shrimp—the only animal that thrives in the saline brew of the lake. It is an ecological axiom that animals able to handle harsh conditions often occur in tremendous numbers. With no competition for food or space, brine shrimp populations in Great Salt Lake are limited only by the amount of algae available to eat. There can be thousands of them wiggling in a gallon of lake water, and swimming among them gives the unnerving sensation of being swarmed by hordes of pink mosquitoes. Breeding and dying in incalculable numbers, the brine shrimp often give the lake a foul odor—and the smell isn't helped by the brine flies, which gather in dense clouds along the lakeshore to feed on dead brine shrimp. The flies and the smell explain why you can often have miles of Great Salt Lake shoreline all to yourself. Stink aside, the brine shrimp form a profitable industry. Their little hard eggs, or cysts, are harvested and sold as fish food worldwide. Great Salt Lake's greatest moneymaker, though, is salt. The salt is harvested from evaporation ponds along the lakeshore and heaped into conical mounds like miniature Mount Fuji's, awaiting shipment. It goes to industrial uses, to making lick-blocks for livestock, and for melting ice on winter roads. If Great Salt Lake were to evaporate entirely, you would see, in its place, a barren, mud-cracked plain coated with glistening salt.

That, in effect, has already happened on the Salt Lake Desert, which spreads its saline crusts across nearly 4,000 square miles of northern Utah west of Great Salt Lake (figure 9.4). As Lake Bonneville dried up, millions of tons of salt crystallized out of its hypersaline waters and fell like snow to the lake bed. As the lake withdrew, the salt stayed behind.

If there is a place on Earth more desolate, inhospitable, and poisonous to life than the Salt Lake Desert, I don't know of it. As a habitat, it is about as close to the Moon as you can get here on Earth. Mark Twain, crossing the desert by stagecoach on his way to Virginia City in August 1861, described it this way:

> Imagine a vast, waveless ocean stricken dead and turned to ashes; imagine this solemn waste tufted with ash-dusted sage-bushes; imagine the lifeless silence and solitude that belong to such a place. . . . The sun beats down with dead, blistering, relentless malignity;

the perspiration is welling from every pore in man and beast; but scarcely a sign of it finds its way to the surface—it is absorbed before it gets there; there is not the faintest breath of air stirring; there is not a merciful shred of cloud in all the brilliant firmament; there is not a living creature visible in any direction whither one searches the blank level that stretches its monotonous miles on every hand; there is not a sound—not a sigh—not a whisper—not a buzz, or whir of wings, or distant pipe of bird—not even a sob from the lost souls that doubtless people that dead air.

Faced with such savage aridity, it's a stretch to imagine a fifth of a mile of water above the Salt Lake Desert. But that's how high up you need to look to see the highest Bonneville shoreline on the flanks of the nearby ranges. Back when waves broke at that level, Lake Bonneville was fresh and teeming with fish. The lake's drying sentenced every resident creature to a slow chemical death, leaving behind salt flats where now only microbes and algae live. Nothing survives on the salt flats that you can see with your naked eye—no plants, no animals, and certainly no people other than visitors who probe the vastness briefly before withdrawing. Hardy plants creep down toward the salt flats from the alluvial fans that skirt the nearby mountains. But at the salt, they stop.

On a bright day—which means most days on the Salt Lake Desert—the salt flats are painfully, blindingly white. Their uninterrupted smoothness reveals the curvature of the Earth. At Bonneville Speedway, on the western margin of the salt flats near Wendover, Utah, racecars seem to sink into the Earth as they drive toward the horizon. During the wet winter and spring months, the speedway floods a few inches deep. In summer, the water dries, and the salt recrystallizes into a hard, flat surface perfect for speed. Serious racing began on the salt flats in 1914, when Teddy Tezlaff drove his Blitzen Benz across the flats at 142 miles per hour. With the world record thus set, daredevils of all types began to converge on a ten-mile swath of salt whose name soon became synonymous with speed. The records rose and fell at Bonneville Speedway: 301 miles per hour by Malcolm Campbell in 1935; 394 miles per hour by John Cobb in 1947; 407 miles per hour by Mickey Thompson in 1960; 526 miles per hour by Craig Breedlove in 1964; 601 miles per hour by Craig Breedlove again one year later; 622 miles per hour by Gary Gabelich in 1970. Gabelich's record stood for thirteen years. By then, premier land speed racing had shifted west, to the table-flat playa of Nevada's Black Rock Desert. (The Black Rock Desert occupies the dried-out bed of Lake Lahontan, a smaller twin of Lake Bonneville that filled much of the western Great Basin during roughly the same time.) There, land speed records took several more leaps forward. Today, the record stands at 763 miles per hour—40 percent faster than a passenger jet—as set by Andy Green, a British Royal Air Force fighter pilot, driving a bullet-shaped car powered by twin turbofan jet engines. Green's car weighed more than ten tons—enough to keep it from taking off, since a land speed record requires that your wheels stay on the ground. To generate the 50,000 pounds of thrust needed for the job, Green's jet engines burned five gallons of fuel per second, for a fuel efficiency of 1/25th of a mile per gallon.

.   .   .

Had Green set his record on the Salt Lake Desert, it would have taken him all of four minutes to cross the salt flats. It took the Donner party five and a half days to cross. Their tracks are still there on the Salt Lake Desert. One can't, of course, point to any given set of faded wagon ruts and claim that the Donner's wagons made them. In the year 1846, the Bryant/Russell, Hoppe, and Harlan/Young parties all crossed the desert ahead of the Donners, and other groups followed during the gold rush years. After the American frontier closed, archeologists and history buffs retraced the Donner's route, deepening the wagon ruts with their own wheels. My presence on the Donner's trail puts me at the end of a long line of rut nuts pulled in by their morbid story—for it was here, on the dried up bed of Lake Bonneville, that their tragedy really began.

The Donner party[3] came down through the forests of the Wasatch Range in late August 1846, nearly one year ahead of the Mormon founders of Salt Lake City. They didn't intend to settle by Great Salt Lake, for they were bound for California. It wasn't gold that drew them there (the gold rush was still three years off), but rather the promise of abundant farmland and a long growing season.

The Donners were following a little-used route to California known as the Hastings Cutoff. Most California-bound emigrants in 1846 followed the established California Trail, which passed well north of Great Salt Lake and the Salt Lake Desert. But the Donners had opted for an alternative route based on the advice of Lansford Warren Hastings, author of a little book called *The Emigrant's Guide to Oregon and California*. Hastings convinced the Donners that they would shave several hundred miles off their journey if they followed his new route, which passed south of Great Salt Lake and directly across the Salt Lake Desert. Hastings assured the Donners that the desert crossing was manageable, taking no more than one day and one night of continuous travel between two springs that lay on opposite edges of the desert. What Hastings had failed to mention was that this was his best guess. The Donners didn't know it, but Hastings had never properly scouted his new route. In fact, at the time he published *The Emigrant's Guide to Oregon and California* in 1845, he had never even laid eyes on it.

Waterless crossings were the hardest challenge that pioneers faced on any overland journey. Water is heavy, and oxen are thirsty, slow-moving beasts—especially when required to haul wagons weighed down with extra water. Look at a map of emigrant trails west, and for the most part you are looking at water routes (the paths of rivers and the locations of springs). Forty miles of waterless travel, although brutal, was manageable (although often at the cost of at least a few oxen, mules, or horses). It could be done by storing as much water and grass as possible in the wagons, and then driving nearly nonstop for a full day and a full night. That was what the Donners anticipated when, on August 28, 1846, they arrived at Hope Wells, the last spring before the Salt Lake Desert.

But at the spring they found a note, left by Hastings several days earlier, that read: "2 days—2 nights—hard driving—cross—desert—reach water." In other words, what lay ahead was a waterless crossing roughly double the distance that the Donners had been led to believe. It was not forty miles but *eighty miles* to the next spring, at the base of the Pilot Range on the opposite side of the Salt Lake Desert.

But there would be no turning back now—not if the Donners intended to get to California that year. "This would be a heavy strain on our cattle," Eliza Donner (a survivor of the cannibalistic winter of 1846–1847) remembered, "and to fit them for the ordeal they were granted thirty-six hours' indulgence near the bubbling waters. Meanwhile, grass was cut and stored, water casks were filled, and rations were prepared" for the eighty-mile dry run.

I picked up the Donner's trail on the crest of the Grayback Hills, where the Salt Lake Desert unfolds in panoramic view to the west (figure 9.5). The snow-white shoulders of the Pilot Range rose sixty miles away, topped by 10,716-foot-high Pilot Peak—the beacon at which the Donners aimed. In between lay an unbroken plain of mud and salt—the residue of Lake Bonneville. The old wagon trail—now a faded Jeep track—dropped down from the hills and disappeared into the void. The Donners rolled forth into that void, their eyes on Pilot Peak and their hopes on the spring at its base. All around them lay an "appalling field of sullen and hoary desolation" in which "the hiatus in the animal and vegetable kingdoms was perfect." They traveled day and night, James Reed remembered, "only stopping to feed and water our teams as long as water and grass lasted." A second day came, and then a second night. And still, all around, lay the implacable desert:

> After two days and two nights of continuous travel, over a waste of alkali and sand, we were still surrounded as far as eye could see by a region of fearful desolation. The supply of feed for our cattle was gone, the water casks were empty, and a pitiless sun was turning its burning rays upon the glaring earth over which we still had to go. (Eliza Donner)

Things began to fall apart. With the water and grass gone, the oxen began to fall over. The men whipped them back to their feet and drove them on. The distances between whippings grew shorter. It became clear that the wagons would have to be abandoned. The men unhitched all of the animals that could still walk and drove them toward Pilot Spring, still more than twenty miles away. As for the people, "all who could walk did so," Eliza Donner recalled, "mothers carrying their babes in their arms, and fathers with weaklings across their shoulders moved slowly as they urged the famished cattle forward." In some cases, children and mothers were left with the wagons on the desert while fathers went ahead to fetch water and return, hopefully, with revived oxen. Children began to wail from fear and thirst, until Eliza's mother "put a flattened bullet in each child's mouth, to engage its attention and help keep the salivary glands in action." A thin line of people and cattle began to stretch west from the abandoned wagons toward Pilot Spring.

Pilot Peak

Silver Island Mountains

salt flats

mud flats

wagon trail

FIGURE 9.5

View west from the top of the Grayback Hills along the eighty-mile waterless stretch of the Hastings Cutoff. This is what the members of the Donner party saw before they descended from these hills to cross the most forbidding section of the Salt Lake Desert. Pilot Spring lies fifty-three miles ahead, on the other side of the Silver Island Mountains below Pilot Peak.

By the time the vanguard of the Donner group had reached the area where I stood to take the top photograph in figure 9.6, the party was strung across the desert like a line of ants. They had now lost some two dozen oxen—many of which, insane with thirst, had broken free and bolted onto the desert, where they collapsed and died. The group was now into the third day of the crossing, and had been without water for more than a day:

> Anguish and dismay now filled all hearts. Husbands bowed their heads, appalled at the situation of their families. Some cursed Hastings . . . for his misrepresentation of the distance across this cruel desert, traversing which had wrought such suffering and loss. Mothers in tearless agony clasped their children to their bosoms, with the old, old, cry, "Father, Thy will, not mine, be done." (Eliza Donner)

As the lead members of the group stumbled to within a few miles of Pilot Spring (bottom photograph in figure 9.6), the oxen sensed water ahead. Lifting their shaggy

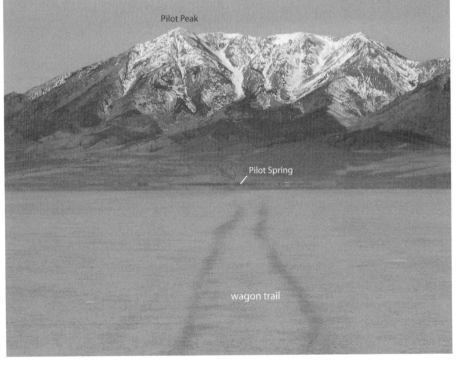

heads, they began to trot. As the salt flats ended, the men whooped, and together, men and cattle crashed through the reeds around the spring and plunged into pools brimming with water percolating down from Pilot Peak.

After resting for several hours and drinking their fill, the men headed back onto the desert with sloshing buckets, driving the revived oxen to carry in other members of the party and to bring in some of the stranded wagons. All told, the back-and-forth rescue-and-retrieval operation took more than two days. (The first wagons, remember, were unhitched more than twenty miles from the spring, so going back to get them added a forty-mile round trip.) Five and a half days after they had left Hope Wells, all eighty-six members of the Donner party had come safely in to Pilot Spring.

The party spent a full week at the spring recuperating, going back for more wagons and jettisoned supplies, and searching for missing cattle. Their losses, when totaled, were staggering. No people had died, but the party had lost thirty-six cattle, which meant that they had to abandon several wagons and hundreds of pounds of crucial supplies. Worse, the entire crossing, including the time spent recuperating at Pilot Spring, had burned up thirteen days. What Hastings had advertised as a one-day-plus-one-night trip took the Donners almost half a month.

Only when the Donners met up again with the original California Trail (along the Humboldt River in what is now northern Nevada) did they realize the scope of their error in following the Hastings Cutoff. Parties that had been *behind* them when the Donners had branched away from the main trail at Fort Bridger, Wyoming, were now in *front*. The calamity would come to a head 800 miles to the west, in late October 1846, when the Donners climbed to within sight of the Sierra Nevada pass that now bears their name, only to be trapped by early season snows. If they had been a few days faster, they would have escaped the blizzards and probably settled into lives of happy obscurity in California. John Breen, one of the survivors, reflecting back on the tragedy in the snows, pointed to the Salt Lake Desert and said, "Here our real hardships commenced."

NOTES

1. Following convention, the age dates that I give throughout this chapter are in carbon-14 years before present, meaning that they are based on the known radioactive decay rate of carbon-14 in organic materials such as shells, bones, teeth, or wood. The dates are close to, but not exactly equal to, calendar years before present. This is because the rate at which cosmic rays produce carbon-14 in the atmosphere varies over time, and therefore the amount

FIGURE 9.6 (OPPOSITE)

The Donner's trail approaching Pilot Spring on the final stretch of the eighty-mile waterless crossing. In the top photograph, I'm looking west along the trace of the wagon ruts where they cross the alluvial fans on the north side of the Silver Island Mountains. The desert plants of the Great Basin grow so slowly that the ruts are still visible. Pilot Spring lies about ten miles ahead. The bottom photograph looks west across the final stretch of mud and salt flats about five miles from Pilot Spring. Often marked as Donner Spring on maps today, Pilot Spring is presently on the property of the TL Bar Ranch.

of carbon-14 added to growing shells, bones, teeth, or wood is not constant. Converting a carbon-14 age to calendar years requires a calibration using tree-ring ages, coral growth banding, or annual layers in lake bed sediments.

2. The Dead Sea and the Caspian Sea are not true seas, but lakes. By geographic definition, seas are bodies of water connected to the ocean, whereas lakes are landlocked.

3. Like most wagon trains, the Donner party consisted of several families, most of them unrelated, along with a number of single men employed as drivers and camp workers. At the time they crossed the Salt Lake Desert, the party consisted of eighty-six men, women, and children. Historians often refer to the group as the Donner/Reed party to acknowledge James Reed's role in its leadership.

10

EVOLUTION'S BIG BANG

In western Utah, a dazzling mountain called the House Range juts nearly a vertical mile up from the sagebrush flats of the Tule Valley. Were such a spectacle to appear anywhere east of the Rockies, it would become an instant national park. Here, though, in this empty quarter of the Great Basin, not even a sign marks it. The closest pavement is twenty-five miles away, the closest gas station eighty miles. You may not see another soul for days here. If you do, the meeting is cause for extended conversation about the roads (bad), the weather (hot), guns (liberals go home), and fossils (for which the House Range is famous).

It's the fossils that I've come for—fossils of animals that lived some 520 to 500 million years ago, during the Cambrian Period of geologic time. In those days, North America straddled the equator, and if I'd been here then, I would have been floating in lukewarm waters above a tropical seabed. The shoreline lay to the east, near Colorado. The Rockies did not yet exist, nor did the Great Basin. What is now Utah and eastern Nevada formed part of a submerged continental shelf that stretched from Colorado west to about the middle of Nevada. There the continental shelf ended, and the deep ocean floor began. The continental bedrock that today makes up western Nevada, along with California, did not yet exist. It would come later, piece by piece, as accreted terranes.

At first, I find no fossils. The oldest rocks that my hammer splits seem barren of life. But farther up through the layers (that is, forward in time), signs of life begin to appear: squiggly burrows about the diameter of a soda straw, made by worm-like creatures as they wriggled across the seabed. Farther on, the burrows become more abundant, and

147

a picture emerges of an ocean floor churning with life. I begin to find pieces of trilobites—crustacean-like arthropods that molted their outer shells, or exoskeletons, as they grew. Then, where I meet the Wheeler Shale and the Marjum Formation, the rocks become marvelously fossil rich. I don't need a hammer now. Fossils—including trilobites, brachiopods, and assorted traces of worm-like creatures—crowd the platy slabs of shale and limestone breaking free from the hillsides. My search through the Cambrian rocks of the House Range has revealed something that European geologists first recognized nearly two centuries ago: the Cambrian Explosion, the big bang of animal evolution. It's a global story—one that reaches far beyond the Great Basin. Yet much of the evidence that reveals it, and the stages leading up to it, has come from the Great Basin, particularly from outcrops in the House Range, Drum Mountains, and Fish Springs Range of western Utah, and the White-Inyo Mountains, Death Valley, and Mojave Desert regions of Southern California.

Throughout the world, wherever marine rocks of the right age are preserved, animal fossils often appear rather suddenly—and sometimes in great abundance—above rocks that seem to be barren of life. This Cambrian Explosion gets its name from the Cambrian Period, which nineteenth-century British geologist Adam Sedgwick defined based on the earliest appearance of shell-bearing fossils, particularly trilobites, in marine strata from Wales. (Cambria was the original Roman name for Wales.) Strata older than Cambrian, which appeared to be devoid of life, came to be known as Pre-Cambrian—a vast and, at that time, poorly known span that represented all of geologic time before the appearance of shell-bearing animals. Presently, radiometric dates and fossil studies have pegged the Cambrian Period as the interval from 542 to 488 million years ago. (I say "presently" because the ages of geologic boundaries can shift in light of new data, although usually not by much.) The Cambrian Explosion occurred roughly in the middle of this interval, from about 525 to 510 million years ago.

The Cambrian Explosion confounded Charles Darwin. In Darwin's conception of evolution, animal fossil diversity should increase steadily through geologic time as evolutionary lineages split into more and more species. Instead—to his puzzlement—huge numbers of animals seemed to appear all at once in Cambrian time, without apparent predecessors:

> To the question why we do not find rich fossiliferous deposits belonging to these assumed earliest periods prior to the Cambrian system, I can give no satisfactory answer. . . . The case at present must remain inexplicable; and may be truly urged as a valid argument against the views [about evolution] here entertained. (*On the Origin of Species*, 1859)

But Darwin then reminded readers, "We should not forget that only a small portion of the world is known with accuracy." Today, a century-and-a-half after *On the Origin of*

Species, we know a lot more about fossils and the history of life. The result is a picture of early animal evolution that would surely have delighted Darwin. His greatest delight, I suspect, might have come from knowing that one of his ideas—a concept we now call coevolution—may explain why the Cambrian Explosion happened.

Presently, we can trace the evolutionary strands of animal life back at least thirty million years before the Cambrian Period, showing that animals were evolving long before Cambrian time, just as Darwin would have predicted. The scarcity of animal fossils in rocks older than Cambrian age arises, in part, from the fact that most Pre-Cambrian animals were soft-bodied and/or small. The fossil record improves dramatically during Cambrian time partly because animals got larger and partly because they started making shells and other hard structures that readily form fossils. This is one reason why animal life appears to "explode" in Cambrian time.

If that was all there was to it, this chapter would be over. But there's more to the Cambrian Explosion than just an improved fossil record. During Cambrian time, evolution ran wild with animal body design, innovating on an unprecedented level. To use a cultural analogy, it's as if Paleolithic humans had invented not just spears but pyramids, catapults, locomotives, and airplanes, too. The result of this Cambrian *explosion of innovation* is that nearly every animal group on Earth today can trace its body design—its "blueprint" for how it is built—directly back to Cambrian time. This arguably makes the Cambrian Period the most important time in the evolution of animal life. It set the course of everything that followed, including the colonization of the land by, among other things, the vertebrates (back-boned animals)—an event that ultimately made it possible for me to write this, and for you to read it.

I'll come back to this point after tracing the evolutionary developments that lead to the Cambrian Explosion, summed up in figure 10.1.

.   .   .

Ediacaran prelude: In 1946, Reginald Sprigg, a young Australian government geologist, idly turned over a slab of sandstone while he sat eating lunch in the Ediacara Hills north of Adelaide. All around, he found delicate, leaf-like impressions of strange-looking animals. Sprigg didn't know it then, but he was peering back to the dawn of visible animal life. Subsequent work showed that this so-called Ediacaran Fauna flourished worldwide from about 570 to 540 million years ago. Ediacaran fossils have now turned up in rocks as far-flung as China, Russia, Namibia, England, Scandinavia, the Yukon, Newfoundland, and assorted sites in the Great Basin, including the White-Inyo Mountains, Death Valley, and Mojave regions (figure 10.2).

Almost all Ediacaran animals were soft-bodied and are preserved as impressions, like handprints, where waves or storms washed sand and mud across their gelatinous corpses. As a rule, soft-bodied creatures are hard to preserve as fossils because scavengers devour

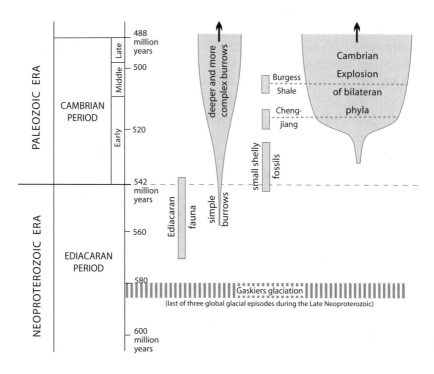

FIGURE 10.1

Major developments in early animal evolution up to and through the Cambrian Explosion.

their corpses, because microbes quickly decompose their soft tissues. But animal predators and scavengers had not yet evolved in Ediacaran days. Moreover, animal grazers had not yet become abundant, allowing mats of cyanobacteria (an ancient form of photosynthetic bacteria) to coat the seabed in sticky layers. The bacterial mats bound together the sand and silt that covered the Ediacaran corpses, preserving the fossils like a death mask. This mode of preserving soft-bodied animals virtually disappeared in later geologic time when grazing animals evolved to eat mats of cyanobacteria, and when scavengers came on the scene.

Most Ediacaran animals look nothing like the animals of later geologic time. When most of us think "animal," we likely visualize a *bilateran*: an animal whose body divides lengthwise into two mirror-image halves. (Picture a person, fish, ant, or flatworm; in each, you can pass an imaginary plane through the body that divides it into two mirror-image halves.) The only nonbilateran animals alive today are sponges, ctenophores, and cnidarians (the group that includes sea anemones, corals, sea pens, and jellyfishes). Bilateran animals have dominated the Earth ever since the Cambrian Explosion. But the Ediacarans, the Earth's earliest animals, were mostly *nonbilateran*. And they were a weird bunch. First off, most of them were immobile. They lacked legs, swimmerets,

FIGURE 10.2
Two examples of the Ediacaran Fauna, a group of animals that flourished worldwide from about 570 to 540 million years ago, before the Cambrian Explosion. Top image: *Dickinsonia,* a headless, flat, pleated creature that lay on the seabed and grew up to five feet long (this specimen is six inches). Bottom image: *Mawsonites,* possibly an early sea anemone, but its evolutionary affiliations are enigmatic (specimen five inches in diameter).

or other features associated with movement, suggesting that they simply sat on the seabed like living lumps. Many of them were round and flat, ranging in size from a penny to a truck tire across. Strikingly, most Ediacaran animals lacked any features that indicate *animal-to-animal interaction.* Nary an eye, antenna, bristle, tooth, claw, or jaw appears among them. Perhaps we shouldn't be surprised—for in a world before predators and prey, what need is there to detect, attack, escape, or hide from other animals? More bizarre, perhaps, is that many Ediacaran animals seem to have lacked any sort of *gut*. How did they eat? Some were pleated like an accordion (top specimen in figure 10.2) and so may have absorbed nutrients directly from seawater using their pleated bodies to maximize surface area. Some may have hosted algae inside their bodies to provide a source of nutrients. (Corals, giant clams, and several other marine animals do the same thing today.)

Nonbilateran, immobile, gutless, and strange, those Ediacaran creatures—*too* strange, it's widely believed, to have been ancestral to most later animal groups. Most Ediacarans appear to represent dead side branches on the animal evolutionary tree—lineages that flourished for a time but left no descendants. Yet there are exceptions. A few Ediacaran fossils look like sea pens or sea anemones (bottom specimen in figure 10.2), and evidence from molecular studies suggests that the cnidarians split from other animals shortly before the Ediacarans appeared.[1] A couple of Ediacaran fossils may represent ancestral mollusks (the group that includes snails, squids, and clams), arthropods (the group that includes crustaceans and insects), or echinoderms (the group that includes sea stars and sand dollars). Perhaps the best way to think about the Ediacaran Fauna is to visualize a once-flourishing evolutionary bush on which most of the twigs, for reasons unknown, die out near the beginning of Cambrian time, with a couple of twigs surviving and possibly contributing to the Cambrian Explosion.

Mobility and biomineralization: While the Ediacaran prelude was in full swing, two developments with great significance for the impending Cambrian Explosion of bilateran animals were unfolding in the wings. The first evidence of animal *mobility* appears about 560 million years ago, in the form of marks made where tiny creatures wriggled across the seabed. Since mobility is a key feature of bilateran animals, many paleontologists think these marks—called trace fossils—represent the first evidence of bilateran life. They take the form of tiny, groove-like burrows, as if someone had doodled on the seafloor with a toothpick. They were made, probably, by some form of primitive worm. The depth, size, and complexity of these seabed burrows increases through Cambrian time (figure 10.1), reflecting an increase in behavioral complexity driven, perhaps, by predators digging into the seabed in search of burrowing prey.

About 547 million years ago, we find the first evidence of biomineralization—the formation of hard body parts like shells and exoskeletons. Biomineralization opened up many new ecological opportunities for early animals, particularly in the realms of predation (claws and jaws for grasping and tearing prey), protection (shells to shield the body from desiccation, temperature changes, and predatory attacks), and movement (attaching muscles to a rigid frame to increase speed and mobility). The earliest shell-makers were so small that they went undiscovered until recently. Collectively, we call them the "small shelly fossils." They include relatives of later Cambrian animals as well as creatures of unknown affinity.

With mobility and biomineralization in place, the stage was set for the Cambrian Explosion.

.   .   .

Back to the House Range, western Utah. Cambrian strata, uplifted with the mountain, jut out everywhere from cliffs, slopes, and canyon walls. A hot wind blows dust off the distant dry lake beds, filling the air with what looks like wood smoke. The wind blows my notes and papers into the sagebrush and fills my eyes with grit. It would be aggravating except for the fossil wealth that spills from the Wheeler Shale, and the case of cold beer in my truck. One reveals the glory of the Cambrian Explosion, and the other helps me to appreciate it.

A hit-and-run collector like me, who spends a day or two fossil hunting in the Wheeler Shale of the House Range, will find mostly hard-shelled creatures—things that readily form fossils, such as trilobites, brachiopods, and other shelly fauna. But the Wheeler Shale also contains a host of soft-bodied animals that more diligent collectors have unearthed over the years. Indeed, the Wheeler Shale contains many of the same animals as the age-equivalent Burgess Shale in southern British Columbia, 800 miles to the north. The Burgess Shale—popularized in bestselling books such as Stephen J. Gould's *Wonderful Life* and Simon Conway-Morris' *The Crucible of Creation*—contains more than 110 species, most of them soft-bodied, giving us perhaps our clearest window onto the Cambrian Explosion. Equally impressive is the slightly older Chengjiang fauna of southern China's Yunnan Province, from which spectacular finds of previously unknown Cambrian creatures continue to emerge.

Collectively, these Cambrian fossil mother lodes reveal a world best appreciated by shedding the constraints of time and plunging with scuba gear into a Cambrian sea. What follows, although fantasy, is based on interpretations by respected experts of the Cambrian Explosion, particularly Simon Conway-Morris. Our Cambrian dive will illustrate two key themes. The first—getting back to a point that I made earlier—is that nearly every animal group on Earth today can trace its body design back to the Cambrian Explosion. The second theme involves predator-prey coevolution, an idea first proposed by Darwin. I will argue that Cambrian seas—once portrayed as mellow ecosystems populated by peaceable grazers—were, in fact, hostile places filled with predators. This is a key point, because the rise of predators may have driven the Cambrian Explosion.

We descend through tropical waters in what, one day, will be the eastern Great Basin, and land on the seabed in a puff of silt. Trilobites scurry from our landing site, leaving miniature bulldozer tracks. Swarming the seafloor all around are hundreds of half-inch-long, trilobite-like animals called *Marrella* (top of figure 10.3). They are like the wildebeests of the Cambrian seabed, more abundant, perhaps, than all the other animals put together. In the trilobites and in *Marrella*, we see the body design of the arthropods, represented today by crustaceans in the sea and insects, chelicerates

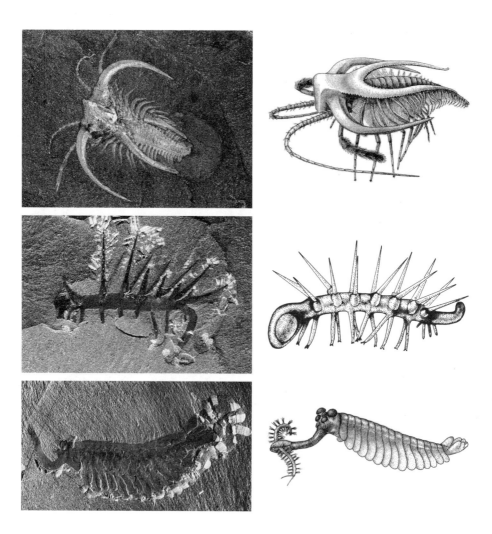

FIGURE 10.3

Selected fossils from the Cambrian Explosion. Top image and ink drawing: *Marrella*, a primitive arthropod with long antennae and an elegant head shield that covered a segmented body (specimen three-quarters of an inch long). Middle: *Hallucigenia*, a creature once thought to be as otherworldly as its name but now thought to be an early lobopod—a phylum represented today by caterpillar-like animals that live in tropical rainforests (specimen three-quarters of an inch long). Bottom: *Opabinia*, an arthropod with five compound eyes on its head and a long, flexible proboscis tipped with a spike-filled claw (specimen 2.5 inches long). The ink drawing shows *Opabinia* clutching a polychaete worm. All images courtesy of the Smithsonian Institution.

(spiders and kin), and myriapods (centipedes and kin) on land. All arthropods have segmented bodies and jointed appendages covered with a suit-of-armor-like exoskeleton that grants them flexibility and protection. This body design—a Cambrian innovation—is clearly a winner, for arthropods today constitute more than three-fourths of all known animal species.

Brachiopods dot the seabed around our landing site. Unable to flee, they snap shut their clam-like shells in our presence. Their similarity to clams is only shell-deep. When we pry one open, we see no fleshy mantle like in clams and other molluscs. Instead, we find a horseshoe-shaped, feathery organ called a lophophore, used for gathering oxygen and food from the sea. The lophophore—another Cambrian innovation—serves brachiopods well to this day.

We kick off and swim slowly across the seabed. Sponges lie scattered about, their spherical or branching forms reminiscent of bushes. They are the simplest of animals, lacking organs or even distinct tissues. They strain seawater for food particles using thousands of flagellated cells, each with a tiny tail that beats like a snapping whip to pump seawater through pores in the body wall. Any sponge that you see today uses the same Cambrian innovation.

Drifting along a few feet above the seabed, we see holes everywhere. Now and then, a column of mud spurts out of one. These are fecal coils, ejected by deposit-feeders—creatures that eat seafloor mud and digest the organic goodies it contains. We plunge our hands into the muck and feel things wriggling away. We catch one and pull it out. The worm's body is segmented, with bundles of bristles called setae projecting from each segment. The bristle-like setae make the worm a very efficient burrower; it uses them for traction as it pushes its way through the mud. The next time you dig up an earthworm, run your fingers along its body; you'll feel a slight roughness. These are its setae, clearly visible under magnification. Your fingers are touching a Cambrian innovation, for setae are the signature feature of the annelids, represented today by earthworms, leeches, and a host of marine polychaete worms.

Plunging my hands into the mud in search of more creatures, suddenly—OWW &#*%#*!! A hideous thing, like a mutant penis, has seized my finger with its circular mouth. It is a priapulid, a unique type of predatory worm with a mouth filled with sharp, backward-pointing hooks. With each gulp, the hooks pull my finger deeper into the priapulid's throat. We can't help but admire the cruel efficiency of this predatory system—once again, a Cambrian innovation that persists to this day.

I cut the priapulid off my finger with a knife. The scent of blood and fresh meat attracts a swarm of visitors. These scavengers, looking a bit like stubby pencils with legs and protective spikes, are called *Hallucigenia* (middle of figure 10.3). They swarm over the cut-up bits of priapulid. When Simon Conway-Morris first interpreted *Hallucigenia,* he mistakenly portrayed it upside down, walking on spike-like legs with a row of tentacles poking up from its back. He picked a name for it suggestive of a psychedelic experience, for *Hallucigenia* seemed so impossibly weird that there was speculation that it might be an entirely new form of animal. But in the 1990s, careful excavation of the shale encasing the fossil by paleontologist Lars Ramsköld showed that the tentacles originally thought to be on the animal's back came in pairs, making them not tentacles but *legs.* Flipped right-side up, *Hallucigenia* became a bit less hallucinogenic. Most paleontologists now classify it as a lobopod, a group of animals represented today by onychophorans—caterpillar-like creatures that live in the leaf litter of tropical rainforests. So here again, a body plan from Cambrian time persists in a group of like-designed animals today.

Scuba divers are always thrilled and spooked to meet a top predator—a creature at the apex of the food chain, like a killer whale or shark. The top predators of the Cambrian seas were close relatives of arthropods. But they bore little resemblance to the meek trilobites and tiny *Marrella* that we saw earlier. A cloud of silt boils up at the edge of our vision. We swim closer and see a worm in its death throes thrashing the seabed. Clutching it is a creature seemingly sprung from the mind of Dr. Seuss. It is *Opabinia*, a predator with a segmented body and a nozzle-like proboscis tipped with a pincher (bottom of figure 10.3). The proboscis protrudes from a head topped with five compound eyes, allowing *Opabinia* to see in practically every direction. Like a faunal version of an AWACS plane, nothing escapes *Opabinia's* notice—including us. Fearful that we'll swipe its catch, *Opabinia* scurries away as we approach, curling its proboscis underneath its head to stuff the worm into its mouth like an elephant feeding itself hay.

Suddenly, *Opabinia* accelerates. We paddle hard in pursuit but can't keep up. What's the alarm? *Opabinia's* keen visual system has detected the threat before we could. A huge beast swims by overhead, zeroing in on the panicked *Opabinia*. It is *Anomalocaris*, which, at five feet long, is by far the largest creature in the Cambrian seas. From the head down, *Anomalocaris* looks much like its intended prey, with a segmented body lined with lateral lobes. But its predatory weaponry is strikingly different. Hanging like fangs from the front of its head are two powerful, jointed appendages that resemble muscular shrimp tails. The appendages are designed to curl back under the body and guide captured prey to the mouth. In hot pursuit of *Opabinia, Anomalocaris* disappears into the Cambrian gloom.

· · ·

Our fantasy dive to a Cambrian seabed showed us a few examples of how the body designs of many animal groups trace back to the Cambrian Explosion—a fifteen-million-year window in time that arguably represents the most innovative period in animal evolution.

The clearest measure of the scope of this innovation comes from taxonomy—the science of sorting organisms into groups linked by common ancestry. Taxonomists have long recognized the special status of groups called phyla (phylum, singular).[2] Animals within a phylum share a common body design. By analogy, picture a Boeing 747, a single-engine Cessna, and a glider. Although quite different in detail, they share a common design: they are fixed-wing aircraft. If they were animals, we'd put them into the same phylum, and we'd put, say, helicopters (a very different aircraft design) into a different phylum. All vertebrate animals possess a backbone with a nerve chord—the common design of the Phylum Chordata. All arthropods have bodies made of hinged, moveable segments protected by an exoskeleton—the common design of the Phylum Arthropoda. All sponges have bodies with specialized cells that pump water through pores in the body wall—the common design of the Phylum Porifera. Each animal phylum represents a distinct body design—so distinct that phyla are easy to tell apart. A spider (Arthropoda) is not built anything like a fish (Chordata), a snail (Mollusca), or a sea urchin (Echinodermata). Phylum-level differences are much greater than apples versus oranges (which, despite the saying, aren't so different, since they are both fruits).

Taxonomists have so far identified about thirty-five animal phyla. Of these, nineteen contain the overwhelming majority of animal species. As figure 10.4 shows, *nearly all, if not all, of these nineteen common animal phyla arose during the Cambrian Explosion.* Arguably, *no* new phyla have evolved since Cambrian time (although there is disagreement about this). Even something as profound for evolution as life moving from the oceans onto land did not involve the production of new phyla. Instead, representatives of Cambrian marine phyla (mostly from the Arthropoda, Mollusca, Annelida, and Chordata) evolved modifications suited for life on land. The Cambrian Explosion ushered in evolutionary innovation on a phylum-level scale never before seen—and unlikely ever to be repeated.

What drove the Cambrian Explosion? Oxygen, it's widely agreed, must have had something to do with it. Microscopic animals can get by on low levels of oxygen, but larger animals—especially if they must grow shells or engage in active, complex behaviors—need more. Did oxygen levels reach a critical threshold during Cambrian time, thus unleashing the Explosion? This hypothesis is hard to test. First, although it's likely that oxygen levels were higher during Cambrian time than before, it's not possible to get exact measurements; all estimates have wide error bars. Second, it not clear what level of oxygen might have been needed to trigger the Explosion. It would have depended, in part, on how efficiently Cambrian animals could extract oxygen from seawater, and that's hard to figure out from fossils. Third, large animals had already evolved millions of years earlier, among the Ediacarans, some of which grew five feet across. Granted, the sedentary life of a typical Ediacaran creature probably required less oxygen than the active, mobile life of a Cambrian animal. But how much less? Given the uncertainties, perhaps the best conclusion to make about oxygen is that although a certain level was doubtless necessary for the Cambrian Explosion, the rise of oxygen alone may not be a sufficient explanation.[3]

Better explanations, perhaps, emerge when we compare Cambrian animals with the Ediacaran animals that preceded them. There are stark differences. The Ediacarans, remember, were mostly nonbilateran. In contrast, the animals of the Cambrian Explosion were nearly all bilateran (figure 10.4). Was there something special about early bilateran animals that allowed them to explode into multiple phyla? In the last several decades, geneticists have discovered that the developmental genes of bilateran animals are remarkably similar across different phyla. For instance, although a mouse (Phylum Chordata) looks nothing a fly (Phylum Arthropoda), the genes that control their early development turn out to be surprisingly alike. In both mouse and fly, very similar genes give

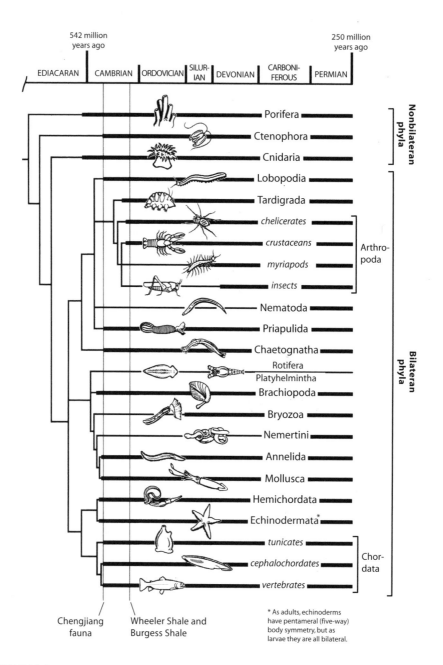

Evolutionary tree of the nineteen most common animal phyla, based on both fossils and molecular similarities among living animal groups. The point where each phylum line thickens marks the time in the geologic record where we find the first fossil representatives of that phylum. Notice that nearly all of the bilateran animal phyla arose during Cambrian time. The three nonbilateran phyla (Porifera, Ctenophora, and Cnidaria) likely diverged from the bilateran animals during the earlier Ediacaran Period. The five phyla not represented by Cambrian fossils (Nematoda, Rotifera, Platyhelmintha, Bryozoa, and Nemertini) are mostly tiny and (except for the Bryozoa) soft-bodied, so their absence in Cambrian rocks may reflect nonpreservation instead of nonevolution.

instructions for making the eyes, limbs, heart, and other parts of the growing embryo. Like switches at a master control board, similar sets of genes direct a mouse along one particular pathway to adulthood, and a fly along another. The point is this: radically different bilateran body designs can spring, apparently, from the simple switching on and off of just a few key developmental genes that all bilateran animals share.[4] But this genetic potential doesn't, by itself, explain the Cambrian Explosion. We need to understand how new body designs, once they appeared in Cambrian seas, were able to survive and prosper. For this, we turn to a complementary hypothesis for the Cambrian Explosion: the hypothesis of coevolution with emerging predators.

To understand the coevolution hypothesis, think back for a moment to the Ediacaran animals. You'll recall that they were a rather dull bunch, entirely lacking features that indicate they interacted with each other. Moreover, most of them were sedentary, soft-bodied, and apparently defenseless. If any predators had been around, the Ediacaran animals would either have been wiped out or, more likely, forced by natural selection to evolve into new forms better able to deal with predatory attacks—in which case, they wouldn't be the Ediacarans that we know. In contrast, most Cambrian animals had features indicative of animal-to-animal interaction—eyes, antennae, jaws, claws, spines, and related features. Moreover, most Cambrian animals were mobile, and many had shells or other hard body parts. Why the stark contrast with the Ediacarans? A compelling hypothesis is that Cambrian animals lived in an ecological arena filled with *predators*. There was *Opabinia,* eyeing the seabed for telltale wiggles and snatching up victims with its nozzle-like proboscis. There was *Anomalocaris,* diving to the seafloor to seize prey with its powerful appendages. There were priapulids lurking in the mud, mouths bristling with fatal hooks. What's a Cambrian creature to do in the face of such threats? The likely answer—as Darwin proposed more than 150 years ago—is to coevolve. If your predator can swim fast, you had better evolve structures that allow you to swim faster or be more agile, for only then will you survive and pass your traits on to your offspring. If your predator has keen eyes, you had better evolve keener eyes or other ways to sense and dodge your assailant. If your predator has powerful jaws or claws, you had better evolve a strong shell or structures and muscles that allow you to burrow, escape, or hide.

Cambrian fossils—unlike earlier forms—show signs of predatory attack. We have trilobites with half-moon–shaped bites missing from their bodies, and brachiopods with holes drilled in their shells. Some Cambrian fossils contain the remains of prey in their guts. We have the arthropod *Sidneyia* with partially digested trilobites, ostracodes, and hyoliths in its stomach, and priapulids preserved with hyoliths and even other priapulids inside. Burrows in Cambrian strata clearly show that animals dug deeper into the seabed over time, possibly to escape predators. A remarkable fossil burrow from Cambrian sandstone in Sweden shows where a predator dug into the seabed to seize a worm. When I think of that fossil, I think of all the creatures on that particular Cambrian day that *didn't* become a predator's dinner, because they were equipped, by evolutionary chance, to dig deeper or faster, thus surviving to pass their abilities on to their offspring.

In sum, the intense and violent selection pressure imposed upon Cambrian animals by predator-prey coevolution may well have driven the Cambrian Explosion. Although Darwin didn't know what we know today about Cambrian fossils, he intuited the immense creative potential of coevolution (a concept he called "mutual relations"). Evolutionary biologists often liken coevolution to an arms race. As predators evolve better weapons for hunting and killing, their prey evolve better ways to escape, hide, or protect themselves, putting pressure back on predators to improve in an endless loop of action and reaction, rewarded or punished by natural selection. Nearly every animal phylum that arose during the Cambrian Explosion shows signs of either predatory ability, the ability to escape predators by burrowing, crawling, or swimming, or the ability to resist predatory attacks with hard shells, plates, or spines. Coevolutionary pressures—when coupled with the inherent ability of the bilateran genome to create variable body designs—may well have spawned the explosion of bilateran animal phyla whose descendants populate the world today.

Those descendants, of course, include us. You and I have a relative from the Cambrian Explosion. It's a rare fossil, however, and it's such a miserable squib of a creature that, had it dwindled to extinction back in Cambrian time, its contemporaries may well not have noticed. Yet if that had happened, you and I might not be here. The earliest-known specimen, found in the Burgess Shale, was first classified as a worm and stuffed into the back of a drawer at the Smithsonian Institution in Washington, D.C. But later studies revealed a notochord—a proto-backbone—indicating that this creature, called *Pikaia*, was a chordate and thus one of the earliest representatives of our phylum, the Chordata. A slightly older version of this animal, called *Cathymyrus*, has since been found in China's Chengjiang beds, pushing the origin of the chordate phylum back to at least that time.

Every backboned animal that has ever lived—every fish and frog, crocodile and dinosaur, chicken and eagle, rhinoceros and grizzly bear—represents a branch on an evolutionary tree that started with *Pikaia* or *Cathymyrus*, or something very like them. Our own human twig on that chordate tree is so recent that it seems almost an afterthought in the pageant of geologic time. Knowing how cognitively advanced we are compared to other animals, it's tempting to think that, somehow, we're the point and purpose of evolution—which, after all, is a process that we—and not frogs, chickens, or *Pikaia*—figured out. But it makes no sense to look for purpose or foresight in evolution. Organisms evolve in response to the immediate demands of their environment, and they do so through the differential reproductive success of variable individuals in their populations. At root, evolution is a simple thing. Yet, as Darwin realized,

from the war of nature, from famine and death, the most exalted object which we are capable of conceiving, namely, the production of the higher animals, directly follows. There is grandeur in this view of life, with its several powers, having been originally breathed into a few forms or into one; and that, whilst this planet has gone cycling on according to the fixed law of gravity, from so simple a beginning endless forms most beautiful and most wonderful have been, and are being, evolved.

NOTES

1. Studies of molecules like DNA, RNA, or the protein sequences of other biochemicals give us a powerful tool, independent of fossil evidence, for deciphering the evolutionary tree of life. The premise is that the more closely related two organisms are, the greater the similarity in their molecular makeup. If we make reasonable assumptions about how fast certain molecules change over time, we can use molecular similarities as a proxy clock—known as the "molecular clock"—to determine, roughly, when the last common ancestor of any two organisms lived.

2. The Swedish botanist Carolus Linnaeus, in 1758, developed the system of taxonomic classification that we use in modified form today. Organisms that can interbreed (or which look so similar as fossils that interbreeding seems likely) are assigned to the same species. Related species are grouped into the same genus (plural, genera), related genera into the same family, and so on, up through ever more inclusive groups called orders, classes, phyla, and kingdoms.

3. For an opposing view, see *Out of Thin Air* by Peter Ward (2006, Joseph Henry Press), a paleontologist who champions the view that oxygen levels were the main driver behind many events in animal evolution, including the Cambrian Explosion.

4. Central to this discovery is the emerging science of "evo-devo," short for "evolution-and-development," whose leading researcher is biologist Sean B. Carroll of the University of Wisconsin at Madison. Carroll's work has shown that the genes of different bilateran phyla are very similar, and that relatively small changes in genetic switches early in an animal's development can create great variability in body form.

PART III

THE ROCKY MOUNTAINS

How I wish I was a geologist, then these rambles over rocks and hills would be of
some benefit to me.

JAMES BERRY BROWN, PIONEER CROSSING
THE ROCKY MOUNTAINS IN 1859

We speak of mountains forming clouds about their tops, [but in fact] the clouds
have formed the mountains. Lift a district of granite, or marble, into their region,
and they gather about it, and hurl their storms against it, beating the rocks into
sands, and then they carry them out into the sea, carving out cañons, gulches and
valleys, and leaving plateaus and mountains embossed on the surface.

JOHN WESLEY POWELL, 1875

Mountains are to the rest of the body of the earth, what violent muscular action is
to the body of man. The muscles and tendons of its anatomy are, in the mountain, brought out with force and convulsive energy, full of expression, passion,
and strength.

JOHN RUSKIN, 1848

11

RANGE-ROVING RIVERS

As mid-nineteenth-century pioneers rolled west in their covered wagons across the Great Plains, they entered a world beyond previous imagination. Most of them had lived their whole lives in the East and the Midwest, on lands that were flat, green, and thick with humus. Now, with each passing day, the land roughened as vigorous streams from the distant Rockies cut sharp valleys into the rising plains. The air, descending from the mountains, lost its ability to condense moisture into rain. With the growing aridity, grasses withered and gave way to sagebrush. The Rockies began to peek up on the horizon. Like gathering clouds, they rose higher each day, casting sunset shadows more than one hundred miles across the plains. "We are now in sight of the highest portion of earth that I ever looked upon," 1852 pioneer William Cornell marveled below 10,270-foot Laramie Peak, in the Laramie Range of present southeastern Wyoming. The Rockies were the first real mountains that most pioneers had ever seen. The mountains they had known in the East and the Midwest, they realized, were mere hills in comparison—bumps dignified with mountain-sounding names. "You may think you have seen mountains and gone over them," forty-niner William Wilson wrote home to his kin in Missouri, "but you never saw anything but a small hill compared to what I have crossed over, and it is said the worst is yet to come."

Passing north and west around the Laramie Range, the pioneers saw thousands of feet of rock strata tilted at improbable angles. Such layers, many realized, had been originally laid down flat. When the deep granite core of the Laramie Range rose, it punched up through the layers and shouldered them aside, so that they now lean against the mountain like lumber stacked against the side of a house. The force and violence of such an upheaval boggled their minds. "This whole region of the world has been an enormous furnace and its power and force incomprehensible," forty-niner Joseph Middleton wrote. What powers, he wondered, could raise these "horizontal rocks into almost perpendicular ridges"? Taking in the mountainous sprawl of the Rockies farther west, 1850 pioneer Lucien Wolcott felt "deeply the inefficiency of my education." "Had I a better knowledge of Geology-Mineralogy-Botany," he noted, "how much more interesting this trip would be?"

Of all the geologic spectacles that pioneers encountered on their trek through the Rockies, perhaps none inspired more commentary than Devils Gate, a 300-foot-deep chasm sawed by the Sweetwater River through a granite ridge in the Sweetwater Hills of south-central Wyoming (figure 11.1). After winding for miles across an open valley, the river barrels straight through the ridge, forming an oasis of green and shade in an otherwise-desiccated landscape of sunburnt ridges and sagebrush plains. "It is grand, it is sublime!" forty-niner John Edwin Banks wrote of Devils Gate, adding, "He must be brainless that can see this unmoved." Lucy Cooke thought it "a grand sight" that was "surely worth the whole distance of travel": "The 'Sweetwater' rushes through an opening in the rocks, the walls on each side rising several hundred feet perpendicularly, and as though riven in two by some great convulsion of nature. . . . I bathed dear little Sarah's feet in the rushing waters, and only wish she had been able to realize the grand occasion."

Yet Devils Gate presented a puzzle. How can a skinny river punch through a 300-foot-high ridge of granite? And *why* would it do so, instead of going around? The river's behavior makes as much sense as driving your car through a stone wall instead of turning to bypass it. Forty-niner A. J. McCall was one of many emigrants who pondered the problem. "It is difficult to account for the river having forced its passage through rocks at this point when a few rods south is an open level plain over which the [wagon] road passes," he wrote. McCall proposed that the ridge "had been rent by an earthquake." Charles Parke thought that the river had indeed gone around the ridge once, but then "there was fire below, but when the fire went out the crust cooled so rapidly, this mountain rib contracted and cracked in twain." Forty-niner Alonzo Delano postulated that the ridge had been split open "by volcanic force," and several other pioneers agreed. Others believed that the river had somehow burst its way through the ridge, perhaps by forcing its way through a fissure. Others credited the work to God. "This is indeed wonderful to look at," Martha Missouri Moore wrote at Devils Gate, "and one stands in awe of Him Who tore asunder the mountains and holds the winds in the hollow of His hands."

The puzzle of Devils Gate is compounded when we realize that it is but one of many rivers in the Rocky Mountains that cut through ridges and uplifts rather than going

FIGURE 11.1

View east downstream along the Sweetwater River through Devil's Gate, south-central Wyoming. The river cuts through this granite ridge even though a clear route around the ridge lies less than one-half mile to the south.

around them. I call them the range-roving rivers of the Rockies. Their paths seem to make no sense. Water flows downhill. How can a river end up going through something that lies uphill of it?

The Wind River flows east for one hundred miles across the Wind River Basin in central Wyoming. Then, for no apparent reason, it angles north and charges at the Owl Creek Mountains, slashing through to carve 2,000-foot-deep Wind River Canyon. Its behavior is so unexpected that early explorers didn't even recognize it as the same river where it emerges on the other side and named it there the Bighorn River. On its way to the Great Plains, the Wind-now-Bighorn River cuts through two more mountain barriers. First, it saws a 700-foot-deep canyon through Sheep Mountain (figure 11.2), even though a clear route around can be found just five miles to the west. Then it cuts across the northern end of the Bighorn Range to carve 800-foot-deep Bighorn Canyon. In southeastern Wyoming, the Laramie River runs north along the west side

FIGURE 11.2

The Bighorn River cutting through Sheep Mountain in northern Wyoming. The view is to the southwest, and the river is flowing toward you. A railroad, which forms the thin line to the right of the river, follows the river through the gap.

of the Laramie Range, seemingly on track to join the North Platte River to go around the range. But instead it turns east and punches through the center of the Laramie Range. The North Platte River cuts through no fewer than three mountain barriers on its way to the Great Plains: the Seminoe Mountains, the eastern Sweetwater Hills, and the Hartville Uplift. Colorado's South Platte River cuts a forty-five-mile-long chasm through the immense Front Range to reach the Great Plains, and in southern Colorado the Arkansas River cuts two deep canyons through the Wet Mountains before debouching onto the plains. The Colorado River is the greatest range-rover of all. From its Colorado headwaters, the river barrels through the towering Park and Gore ranges as it heads toward the Colorado Plateau. There the river goes through numerous ridges and uplifts, including the Kaibab Uplift to carve the Grand Canyon. Bottom line: it's hard to find a river *anywhere* in the Rocky Mountains or the Colorado Plateau that does not cut through a large rock obstacle somewhere along its route. The crowded field labeled on figure 11.3 is just a partial inventory for the Rockies, and it doesn't include the many examples from the Colorado Plateau.

How have these rivers come to cut through the mountains? The answer appears to be that the rivers once flowed *above* the mountains and have since cut down through

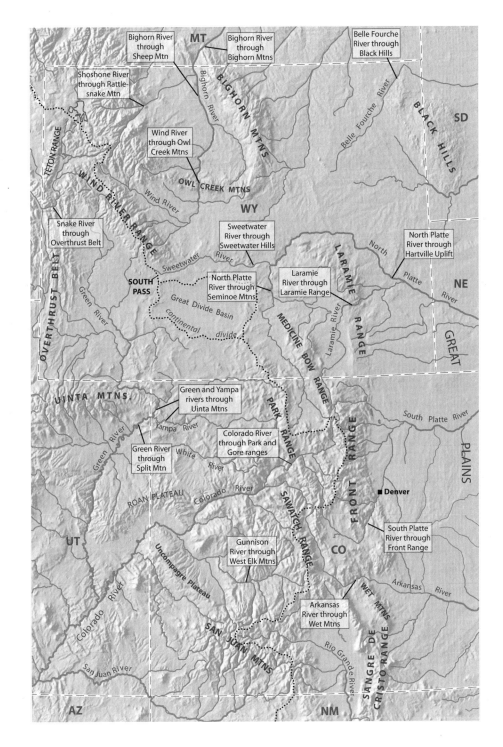

FIGURE 11.3

A partial list of Rocky Mountain rivers that cut through ranges, ridges, and uplifts instead of going around them.

them to establish the paths that we see today. But how did that happen? As we'll see shortly, the evidence shows that the Rockies once lay *buried* under thick layers of sand and gravel, with only the highest peaks poking out. Rivers flowed across these gravelly plains, passing over the entombed mountains. Then, invigorated by uplift of the region or by wetter climates (a debate that we'll explore in chapter 13), the rivers began to cut downward, using for teeth the sand and gravel that they carried in their beds. The rivers bulldozed away the loose gravel layers above the mountains and sawed on through the mountains beneath. The result is the rugged, river-slashed Rocky Mountain landscape of today—a landscape shaped less by the original *uplift* of the Rockies than by their subsequent *burial and exhumation*. Geologists call this momentous event the Exhumation of the Rocky Mountains, or simply, the Exhumation.

Big insights in science often arise out of small problems. By probing at a narrow problem—how Rocky Mountain rivers came to cut through the mountains—geologists dug up one of the biggest stories in the geologic history of the American West: that of the Exhumation.

Our journey to this insight begins with nineteenth-century geologist and explorer John Wesley Powell. Son of an abolitionist father, Powell fought in the Civil War against those who would preserve slavery, losing his right arm to a rebel bullet at the battle of Shiloh in 1862. To be so crippled in the prime of life (he was twenty-eight at the time) might have dampened the ambitions of most men, but for Powell it seems almost to have had the opposite effect. After serving out the war and rising to the rank of major, Powell set out west to explore and map the Rocky Mountains. His work there set the stage for a grander ambition—to explore, by river, the unknown canyons of the southern Rockies and Colorado Plateau, including the Grand Canyon.

On May 24, 1869, Powell set out with nine companions in four rowboats, launching onto the Green River from the tiny town of Green River in what is now southern Wyoming. Over the next three months, the Powell Expedition would travel nearly 1,000 miles down the unexplored Green and Colorado rivers, through winding chasms and tumultuous whitewater rapids, mapping and surveying vast tracts of canyon wilderness in what is now Utah and Arizona. As the townspeople turned out to watch the expedition launch, Powell wrote in his notebook:

> We take with us rations deemed sufficient to last ten months, for we expect, when winter comes on and the river is filled with ice, to lie over at some point until spring arrives; and so we take with us abundant supplies of clothing, likewise. We have also a large quantity of ammunition and two or three dozen traps. For the purpose of building cabins, repairing boats, and meeting other exigencies, we are supplied with axes, hammers, saws, augers, and other tools, and a quantity of nails and screws. For scientific work, we have two sextants, four chronometers, a number of barometers, thermometers, compasses, and other instruments.

Powell suspected that the spring-engorged Green River would cause trouble for the expedition where it pinched through the Uinta Mountains several days' travel downstream. "The river is running to the south; the [Uinta] mountains have an easterly and westerly trend directly athwart its course, yet it glides on in a quiet way as if it thought a mountain range no formidable obstruction," he observed as the boats floated closer to the mountains. "It would seem very strange that the river should cut through the mountains," Powell later wrote, "when, apparently, it might have passed around them to the east."

At first, it appears that the Green will do that—pass around the Uinta Mountains to the east. After barreling into the range's ruddy foothills to cut "a flaring, brilliant red gorge" that Powell named Flaming Gorge, the river turns east to follow the northern flank of the Uintas for thirty miles through a wide valley called Brown's Park. But then, in defiance of topography and expectation, the river turns south and drives directly into, and through, the Pre-Cambrian core of the mountains, cutting 2,500-foot-deep Lodore Canyon (figure 11.4, location 1).

Powell and his men didn't have much time to ponder this turn of events. As soon as the river plunged into Lodore's gloomy depths, the brown Green turned white. Where rivers slash through mountains, they make rapids, both because the canyons pinch the channel and because side canyons dump rocks into the riverbed during rainstorms. With no experience in running whitewater, Powell's men had to learn by doing—which, as often as not, meant capsizing. When "the boat is capsized, then we must cling to her," Powell explained, "for the water-tight compartments act as buoys, and she cannot sink; and so we go, dragging through the waves, until still waters are reached, when we right the boat and climb aboard." On the first day in Lodore Canyon, one of the boats struck a large rock and snapped in two. No one was hurt, but it was a sobering loss so early in the trip. Happily, they caught up with the wreckage downstream and were able to rescue the barometers (essential for their surveying work) and—to the men's greater delight—a large keg of whiskey.

Several days' travel down Lodore Canyon, a new development compounded Powell's puzzlement about how the Green River came to cut through the Uinta Mountains. Another river, the Yampa, joins the Green from the east (figure 11.4, location 2). It does so in the center of the mountain, near the deepest part of Lodore Canyon. It was odd enough to have a river cut through a mountain—but to have *another* river join it in the center seemed stranger still.

In due course, the Green emerges from Lodore Canyon to flow across an open valley south of the Uinta Mountains. Ahead to the southwest, a colossal ridge of bowed-up rock layers rises 2,600 feet above river level. The ridge slopes down to the west, however, and merges with the surrounding valley floor, thus providing a route for the river around it. But instead of going that way, the river drives straight through the ridge and splits it in two. Powell dubbed the ridge Split Mountain (figure 11.4, location 3; also figure 11.5).

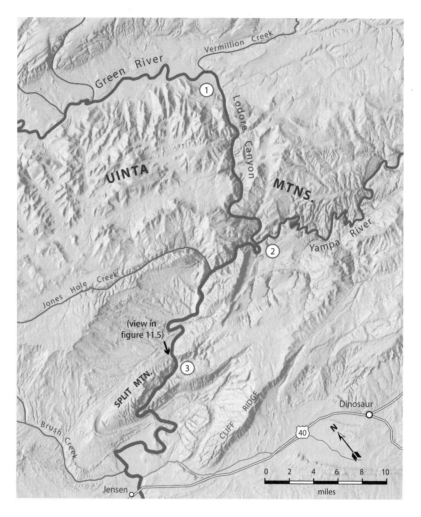

FIGURE 11.4

The Green River where it cuts through the Uinta Mountains near the Utah-Colorado border. The river does three unexpected things here, labeled 1, 2, and 3 on the map. Location 1: Instead of continuing east through low country, the river angles south and cuts through the Uinta Mountains to make Lodore Canyon. Location 2: The Yampa River, after cutting its own canyon through the mountains, joins the Green River in the deepest part of Lodore Canyon. Location 3: After exiting the Uinta Mountains, the Green River then cuts through Split Mountain instead of going around it. See figure 11.5 for closer views of Split Mountain.

By now, Powell was thoroughly perplexed. The Green and Yampa rivers, he realized, were behaving as if the mountains *were not there*. Eventually (his writings don't reveal exactly when), an explanation dawned on him. Perhaps, he decided, the rivers were flowing in their current paths before the mountains rose.

The Uinta Mountains and Split Mountain, Powell could see, were made of up-folded rock strata, similar to the folds made if you scrunched together a flat-lying stack of tow-

This morning we enter Split Mountain Canyon, sailing in through a broad, flaring, brilliant gateway.
—John Wesley Powell, June 25, 1869

FIGURE 11.5

Split Mountain on the Green River in northeastern Utah. The top image looks obliquely southwest over Split Mountain. (From Google Earth: Image USDA Farm Service Agency, © 2010 Google, Image © 2010 DigitalGlobe, Image State of Utah.) The river, flowing away from you, goes through the mountain even though an unobstructed way around lies a short distance to the west (to the upper right). The bottom photograph looks downstream from river level where the river enters Split Mountain.

els on a table. Perhaps, Powell reasoned, the Green River had been flowing in its current path before the rock layers arched upward to form the mountains. "The river . . . was running ere the mountains were formed," he proposed, and "as the fold was lifted, it cleared away the obstruction by cutting a cañon." Clearly, the uplift of the mountains must have been slow enough for the river to cut down apace. "The emergence of the fold above the general surface of the country was little or no faster than the progress of the corrasion of the channel." Powell called his theory antecedence, because it held that the paths of the rivers predated (were antecedent to) the rise of the mountains.

Voilá! It appeared that the mystery of the range-roving rivers of the Rockies might be solved. All that remained was to show that the paths of the rivers predated the rise of the mountains. In Powell's mind, this was practically a foregone conclusion. Like many geologists of his day, Powell believed that mountains were youthful features of the Earth's surface—so youthful that many of them are still rising today, albeit very slowly. If mountains are young, then it makes sense that the paths of large, established rivers might predate their uplift. As we'll see shortly, Powell was wrong about the youthfulness of the Rockies. The mountains are actually much older than the rivers that cut through them, so Powell's antecedent theory can't be right. But his mistake teaches an interesting lesson about the history of geologic thought—one that's worth an aside before we come back to solve the problem of the range-roving rivers.

.   .   .

Ever since the birth of our profession, geologists have grappled with two fundamental questions: How did the ocean basins form, and why do belts of mountains exist on the continents? Today, we're pretty sure that plate tectonic theory gives us the answers. Ocean basins form where continental plates tear apart and the seafloor spreads in between. Mountain belts form where oceanic plates subduct under continents, or where a continent collides with an arc of islands, or where two continents smash into each other.

In Powell's day—a century before the discovery of plate tectonics—geologists had a different theory for the origin of ocean basins and mountain belts. They called it thermal contraction theory. According to contraction theory, the Earth has shrunk as it has cooled over time. The ocean basins represent places where huge slabs of the Earth's crust have collapsed into its shrinking interior, like the down-dropped sections of an old sidewalk. The continents represent slabs of crust that have stayed high. With continued contraction, the slabs that form the ocean basins have pushed sideways against the continents, wrinkling their edges to form mountain belts. Just as the skin of a drying apple wrinkles as it shrinks, so mountains have arisen on the skin of the shrinking Earth.

Powell figured that since the Earth is still cooling and shrinking, mountains must still be rising on its shriveling skin. "The contraction or shriveling of the Earth causes the rocks near the surface to wrinkle or fold," he asserted, and it so happened that "such a fold was started athwart the course of the [Green] river. . . . The river preserved its level, but the mountains were lifted up; as the saw revolves on a fixed pivot, while the log through which it cuts is moved along. The river was the saw which cut the mountains in two."

Contraction theory turned out to be wrong. Its demise arose from two discoveries around the turn of the twentieth century. As geologists began making detailed maps of the folded-up rock layers in mountain belts like the Alps and the Appalachians, they ran into a problem. When they "unfolded" the rocks geometrically, like straightening out a stack of rumpled towels, they found that there were unexpectedly large amounts of contraction to account for—hundreds of miles worth. Added up over all the mountain belts of the world, the amount of contraction demanded by the rocks vastly exceeded any reasonable calculations of the amount that the Earth has likely shrunk as it has cooled. Shortly thereafter, physicists discovered radiogenic heat—the production of heat inside the Earth from radioactive decay.[1] Radiogenic heat implied that the Earth was not cooling down as fast as previously assumed. In short, the folded-up rocks in the mountains demanded *more* contraction than seemed reasonable, yet radiogenic heat implied there had been *less* contraction than expected.

Faced with this one-two punch, contraction theory lost all credibility. Geologists, after wandering in the intellectual wilderness for a time, then rallied around another explanation for mountain belts called geosynclinal theory. This theory held that mountain belts begin where sediments pile up thickly in great troughs, called geosynclines, which form under the oceans alongside the continents. Eventually, according to the theory, the sediments sank deeply enough under their own weight that they would become heated, distorted, and intruded by molten rock. Then—for reasons that geosynclinal theory never adequately explained—the whole distorted mass somehow rose up into mountain belts along the edges of the continents.

Frustratingly vague on the very thing it set out explain, geosynclinal theory nonetheless persisted well into the twentieth century, in part because geologists had no better ideas about where mountain belts come from. A better idea came tantalizingly close in 1915, when the German scientist Alfred Wegener proposed that the continents had drifted apart to open the Atlantic Ocean. But the idea seemed too fanciful and was, for the most part, hotly rejected. Geologists didn't sit on their hands during this period of theoretical limbo: they mapped vast regions of the Earth's surface, cataloged and classified rocks and fossils, and discovered many of the petroleum and mineral reserves that have lifted society into the modern age. But it was a frustrating time because the science lacked a unifying theory on which to hang all of its data. That began to change during and after World War II, when sonar surveys of the ocean floor revealed, for the first time, the worldwide distribution of mid-ocean ridges and ocean trenches. These discoveries precipitated the plate tectonic revolution in the 1960s. The discovery of plate tectonics kick-started the profession anew, giving it a robust theoretical foundation for understanding—among other things—where mountain belts come from.[2]

For John Wesley Powell, the contraction theory of mountains led logically to his ante-cedent theory for how rivers came to cut through the Rockies. Contraction theory im-plied that mountains are young, still-forming features of the Earth's surface, so it logi-cally followed that preestablished rivers could cut through young mountains as the mountains rose. But in the same 1875 report where Powell introduced his antecedent theory, he gave ample space to an alternative theory of range-roving rivers—one cham-pioned in 1874 by Rocky Mountain geologist Archibald R. Marvine.

According to Marvine, the Rocky Mountains rose long before the present paths of the rivers that now cut through them. The mountains then gradually eroded, filling up the intervening valleys with sedimentary debris until all but the highest peaks were covered over with smooth, gravelly plains. Rivers crisscrossed these plains un-affected by the buried mountains below, often by chance passing right over them. Then the rivers gradually cut down through the cloak of sedimentary layers and, in Marvine's words, "commenced sinking their cañons into the underlying complicated rocks" beneath.

In his wanderings through the Rockies, Marvine had noticed something curious: flat-lying benches and plateaus at high elevations, typically between 9,000 and 11,000 feet above sea level. He wondered if these flat surfaces represented the fill level of sedi-mentary layers that had "once extended up over what are now the mountain rocks." If so, he proposed, "it is but recently that the upper rocks have been completely removed from the summits of the mountain spurs," and the evidence lay in "the often uniform level of the spurs and hill-tops over considerable areas, and large plateau-like regions which become very marked from certain points of view."

Powell, who had a talent for naming things, called Marvine's idea superimposition, because it held that rivers superimposed themselves onto buried mountains. Wanting, apparently, to have it both ways, Powell wrote, "I fully concur with Mr. Marvine in the above explanation of the valleys of the main Rocky Mountains of Colorado." But he then insisted that his antecedent theory was the correct explanation for the canyons cut by the Green River through the Uinta Mountains and Split Mountain, and for those cut by the Colorado River across the Colorado Plateau.

Any scientific theory must make testable predictions. If Powell's antecedent theory is correct, the evidence must show that the path of a river is older than the uplift of the mountain through which it cuts. If Marvine's superimposition theory is correct, the evi-dence must show that the mountains were once covered over with sedimentary layers and then uncovered as the rivers cut down into the buried mountains below. How, you may wonder, can we figure out the age of a river or date the rise of a mountain? The an-swer is that we look for the oldest sediments laid down by the river or shed from the rising mountain. If we can figure out the age of those sediments (from the fossils they contain or from interbedded layers of lava or volcanic ash that we can age-date using radioactive elements), then we can tell when the river began to flow in its current path and compare that to when the mountain began to rise.

Geologists after Powell and Marvine showed conclusively that the ranges and uplifts of the Rockies, and the Colorado Plateau, are much *older* than the present paths of the rivers that cut through them. The ranges and uplifts, we now know, date to the Laramide Orogeny—that momentous mountain-building episode (introduced in chapter 6) that shoved up the Rockies and warped the Colorado Plateau between about eighty million and forty-five million years ago. The present paths of Rocky Mountain rivers turn out to be far younger than this.

Meanwhile, geologists explored and mapped the "large plateau-like regions" that Archibald Marvine had spotted at high elevations throughout the Rockies. They realized that Marvine was correct; these flat-lying regions on the high mountainsides appeared to mark the fill level of sand and gravel debris that had once covered all but the highest Rocky Mountain peaks. Archibald Marvine—without naming it such—had discovered the Exhumation of the Rocky Mountains. Today, we call these high-elevation surfaces *sub-summit surfaces* because they occur at elevations a little below the highest Rocky Mountain summits. They look like broad steps, several miles wide, notched into the mountainsides between about 9,000 and 11,000 feet. Figure 11.6 shows an example.

Many Rocky Mountain geologists think that at least some of these sub-summit surfaces represent abandoned pediments. Pediments are smooth, nearly flat surfaces cut by streams onto bedrock where mountain fronts retreat laterally from their adjacent valleys. Normally, pediments merge smoothly with the surfaces of adjacent, gravel-filled valleys. But in the Rockies, the sub-summit surfaces end in thin air thousands of feet above today's valley floors. Refill, in your mind's eye, the valleys of the Rocky Mountains up to the level of the sub-summit surfaces. You look then onto a Miocene world, some five to ten million years ago, when only the highest peaks poked like islands above deep seas of sand and gravel. Today's Trans-Antarctic Mountains—covered by more than a mile of ice with only the highest ridges exposed—present a comparable image. Figure 11.7 sums up the story of the Exhumation of the Rocky Mountains, as we presently understand it.

As often happens in science, the answer to one question—in this case, Why do Rocky Mountain rivers cut through mountain ranges instead of going around them?—brings up a host of new questions. Why were the Rockies uplifted? Why were they then buried? And why didn't they *stay* buried; in other words, why did the Exhumation happen? We'll explore the answers, and the evidence, in the next two chapters.

.   .   .

From Split Mountain, Powell and his men drifted south down the Green River to its confluence with the Colorado River in what is now Canyonlands National Park. From there, they continued south into the Grand Canyon, which they entered during the second week of August 1869, some twelve weeks after setting out. By now, the intellectual puzzle of how rivers can cut through mountains had taken a back seat to survival. The men were now more than 600 miles into unexplored territory, with no idea of how many more miles of canyons lay ahead. The expedition had negotiated so many rapids

Labels within the image:
- Dubois
- 287
- Lander
- WIND RIVER RANGE
- 28
- sub-summit surface
- 0   10   20
- miles
- N
- sub-summit surface (elevation ~10,000 feet)
- Valley carved by glacier after sub-summit surface formed.

FIGURE 11.6

The main sub-summit surface on Wyoming's Wind River Range forms a ten-mile-wide eroded plateau on the range's southwest flank (top). The sub-summit surface lies at an elevation of about 10,000 feet, several thousand feet higher than the valley to the south and several thousand feet lower than the highest peaks. The photograph (bottom) looks north from the air at a portion of this sub-summit surface near the Green River Lakes. The once-continuous surface is now deeply cleaved with valleys cut by Ice Age glaciers.

that the men had lost count. Swampings and capsizings continued, and the time taken to retrieve and dry supplies, and to make repairs on damaged boats, was a huge drag on progress. The work and time required to portage the rapids hardly made that a better option—yet Powell insisted on frequent portages, where practicable, since every rapid held the potential for tragedy. "Every waking hour passed in the Grand Canyon has been one of toil," Powell wrote. "We have watched with deep solicitude the disappearance of our scant supply of rations, and from time to time have seen the river snatch a portion of the little left, while we were a-hungered."

Frank Goodman had been the first to lose heart; he had walked away from the expedition soon after Split Mountain. William Dunn and two brothers, O. G. Howland and Seneca Howland, abandoned the expedition in the Grand Canyon on August 28, 1869, at what is now called Separation Rapids. "They entreat us not to go on," Powell wrote of the three, "and tell us that it is madness to set out in this place; that we can never get safely through." Just one day after those three men left the expedition to try to escape the Grand Canyon on foot, Powell and his five remaining companions floated out of the canyon through the Grand Wash Cliffs, and soon found themselves in the safety of a Mormon settlement. They returned to civilization several months later, to be hailed as heroes and triumphant explorers. None of the four men who abandoned the expedition was ever heard from again.

Powell, we now know, was wrong about the antecedent origin of range-roving rivers in the Rocky Mountains and the Colorado Plateau. But he wasn't entirely wrong about antecedent rivers. Such rivers do exist. Near Santa Barbara in California, a mountainous bulge called the Ventura Anticline rises on the land like a cat arching its back. The bulge is clearly younger than the Ventura River that cuts through it. The river cuts its channel apace with the rising bulge, making Ventura Canyon. In the Himalaya, the Kosi, Arun, and Tsangpo rivers begin north of the high Himalayan Front, cutting spectacular gorges through the range crest as they flow south to the Ganges River. Deposits left by these rivers on the Ganges Plain suggest that they are older than the mountains. Some parts of the Himalaya are rising at the blistering pace of nearly one inch per year. Yet the rivers—their turbid channels packed with abrasive, tumbling rocks—keep pace, cutting down as the mountains rise around them.

Although I know of no evidence to prove it, I suspect that Powell's antecedent theory may explain some of the range-roving rivers of the Basin and Range, where the mountains are much younger than the Rockies. For instance, Nevada's Humboldt River slashes through the Tuscarora Mountains and the Adobe Range to form Palisade Canyon and Carlin Canyon (home to the outcrop shown in figure A.1 in appendix I). Superimposition won't work as an explanation for these canyons; the mountains are too young and have never been buried. Picture the ancestral Humboldt meandering west across a relatively flat pre–Basin and Range landscape. As the crust stretched and broke up, the river may have maintained its course through some of the rising ranges. In other places,

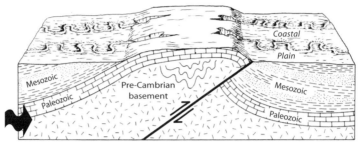

Coastal Plain

Mesozoic

Paleozoic

Pre-Cambrian basement

Mesozoic

Paleozoic

COMPRESSION

~75 million years ago

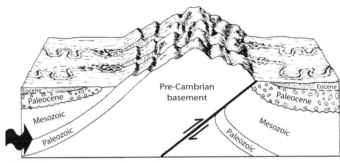

Eocene

Paleocene

Mesozoic

Paleozoic

Pre-Cambrian basement

Eocene

Paleocene

Mesozoic

Paleozoic

COMPRESSION

~50 million years ago

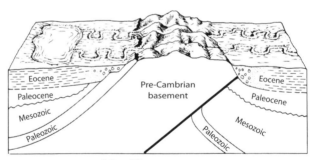

Eocene

Paleocene

Mesozoic

Paleozoic

Pre-Cambrian basement

Eocene

Paleocene

Mesozoic

Paleozoic

~35 million years ago

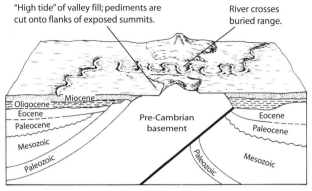

"High tide" of valley fill; pediments are cut onto flanks of exposed summits.

River crosses buried range.

Oligocene
Miocene
Eocene
Paleocene
Mesozoic
Paleozoic

Pre-Cambrian basement

Paleozoic

Eocene
Paleocene
Mesozoic

~ 5–10 million years ago

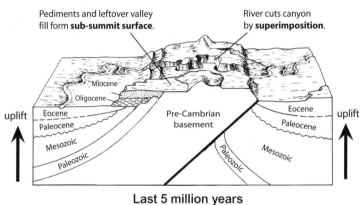

Pediments and leftover valley fill form **sub-summit surface.**

River cuts canyon by **superimposition**.

Miocene
Oligocene

uplift

Eocene
Paleocene
Mesozoic
Paleozoic

Pre-Cambrian basement

Paleozoic

Eocene
Paleocene
Mesozoic

uplift

Last 5 million years

FIGURE 11.7

The probable sequence of events that produced both the range-roving rivers and the sub-summit surfaces of the Rocky Mountains. The Rockies began to rise about eighty million years ago due to massive sideways squeezing and movements along big, ramp-like faults called thrust faults—faults in which the rock above the fault is thrust up and over the rock below. Erosion stripped rock debris from the up-bulging mountains and swept it out into the adjacent valleys. The ranges stop rising about forty-five million years ago, but erosion kept cutting down the mountains and filling up the valleys. Contributing to the burial were thick layers of wind-blown volcanic ash coming, in part, from the great caldera eruptions of the future Great Basin to the west. By about ten million to five million years ago, the sedimentary and volcanic layers had piled up so thickly that they had covered many of the lower ranges, leaving only the highest peaks exposed. Rivers established paths across this landscape, often by chance passing over the buried mountains. Meanwhile, pediments were notched around the exposed summits at the maximum level of valley fill. Later, the region began to be uplifted (for reasons that we'll explore in chapter 13), prompting the rivers to slice downward into the once-buried mountains via superimposition. With uplift, the pediments were raised to their current elevations (9,000 to 11,000 feet) to become today's sub-summit surfaces. (Reproduced with permission of the Wyoming State Geological Survey.)

the ranges may have risen too fast for the Humboldt to keep up. The rising mountains would have shunted the river aside, sending it onto routes around their ends. Thus, we have today's Humboldt River, cutting through mountains in some places and passing around their ends in others. Data may one day sully this story, but until then, I'll stick by it—and raise a glass to John Wesley Powell: soldier, explorer, and one-armed geologist extraordinaire of the nineteenth century.

NOTES

1. Radioactive forms of the elements potassium, uranium, and thorium in the mantle are the main sources of the Earth's radiogenic heat. As I explain in appendix I, radioactivity also provides a natural clock that allows us to calculate the age of rocks and estimate the age of the Earth.

2. This brief recounting doesn't do justice to the story of plate tectonics and its discovery. For a good general summary, see McPhee 1998 (116–137). For an in-depth account told by the scientists who shared in the discovery, see Oreskes 2001.

12

UP FROM THE BASEMENT

Towns in Colorado and Wyoming often post their elevations, and sometimes their populations, on signs at the outskirts. It's a curious tradition. Is it to help arriving tourists break the ice with locals? "I see that your town is 7,362 feet above sea level—that's some thin air!" More likely, it reflects the commanding influence of geography on the lives of Rocky Mountain residents. Colorado and Wyoming encompass some of the highest and emptiest regions in the United States, and the elevations of many towns thus far outstrip their populations. Allenspark, Colorado: 8,450 feet above sea level, population 496; Centennial, Wyoming: 8,076 feet, population 192; Alma, Colorado: 10,578 feet, population 179; Buford, Wyoming: 8,000 feet, population 2. Denver may be the Mile-High City, but its location—at the foot of the Rockies rather than in them—puts it at a lower elevation than most Colorado or Wyoming towns.

Colorado and Wyoming are America's highest states, averaging 6,800 feet and 6,700 feet above the sea, respectively. Utah comes in third at 6,100 feet, New Mexico, Nevada, and Idaho each break 5,000 feet, and the rest of the field is hardly worth mentioning. At 3,400 feet, misnamed Montana (two-thirds of which is Great Plains), is only half as high as Colorado, and Alaska, despite having the highest peaks, is even further down the list at 1,900 feet.[1] Colorado's elevation credentials outrank every other comparably sized space in North America. The state has more fourteeners—peaks higher than 14,000 feet—than all other U.S. states combined, and more than all of Canada too. Colorado's lowest point (3,315 feet above the sea along the Kansas border) is higher than the *highest* point in twenty other states. Colorado's loft puts it at number·

one among U.S. states in the per capita death rate from lightning strikes. Colorado and Wyoming are, in effect, the lumpy roofline of the lower forty-eight states. Rivers begin here and flow away to all points of the compass. Colorado receives no rivers from another state (unless you count the Green River's brief in-and-out foray across the border from Utah; see figure 11.3). Wyoming receives one river, the North Platte, from the only state uphill of it—Colorado—and sends rivers to all other neighboring states. Wyoming's Wind River Range is the only mountain in North America that supplies water to all three master streams of the American West: the Missouri, Colorado, and Columbia rivers.

Lofty Colorado and Wyoming are home to what may be the nation's most magnificent mountains—a collection of skyscraping hulks known to geologists as the Rocky Mountain Foreland Ranges (figures 12.1 and 12.2). The Foreland Ranges represent a special subset of the Rockies. They include southern Montana's Beartooth Mountains, Wyoming's Bighorn, Wind River, Owl Creek, Sierra Madre, Medicine Bow, and Laramie ranges, South Dakota's Black Hills, Colorado's colossal Front Range, along with the Park, Sawatch, and Sangre de Cristo ranges (the last of which pokes south from Colorado into New Mexico), and Utah's Uinta Mountains. When most of us in the United States picture the Rocky Mountains, the Foreland Ranges are what come to mind.[2]

Until recently, the Foreland Ranges were an unsolved puzzle in the geologic story of the American West. The great upheaval that raised the ranges came to be known as the Laramide Orogeny (named for Wyoming's Laramie Range), but putting a name on the event didn't shed much light on it. The Laramide Orogeny presented geologists with two vexing problems. First, how did mountains arise so far in the interior of the continent? And second, how did large blocks of deep, ancient rock—rock known as continental basement—rise from miles down in the crust to make these mountains?

Today, we think that plate tectonic theory has given us answers to both questions—although in a way that early purveyors of the theory would never have guessed. You may remember that back in chapter 6 I gave away the tectonic punch line about how the Foreland Ranges probably formed: via flat subduction of the Farallon Plate. But I didn't do justice to the story there; I fed you the frosting of an idea without the underlying cake of evidence that supports it. Here, I want to show you the geology of the Foreland Ranges and lay out the evidence for how we think they formed.

You can see on figure 12.1 that the Foreland Ranges lie farther inland than any other part of the North American Cordillera. They seem to occupy a space where the Great Plains ought to be, bulging east from the rest of the Cordillera like a salient in a line of battle. In your mind's eye, take away the Foreland Ranges from figure 12.1. In this alternative geography, the Cordillera would reach no farther east than Utah, and Colorado and Wyoming would be part of the Great Plains.

The interior location of the Foreland Ranges presents a problem. Plate tectonic theory tells us that most mountain belts arise along the *edges* of continents as byproducts of subduction. Picture, for instance, the Andes, lofted where the Nazca Plate plunges

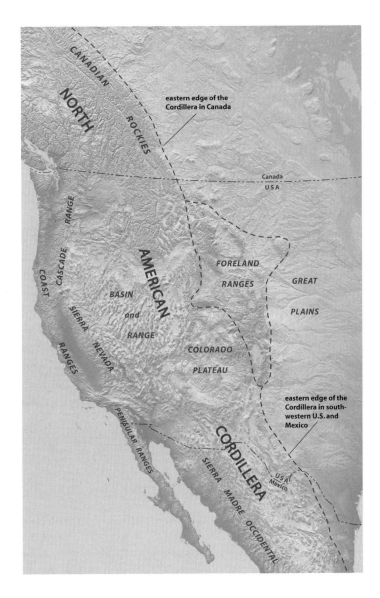

FIGURE 12.1

Location of the Rocky Mountain Foreland Ranges in the North American Cordillera. Notice that the ranges project inland from the rest of the Cordillera, occupying a space where you might expect the Great Plains to be.

into the Peru-Chile Trench. Lay a ruler on a map, and you'll see that the Andes typically crest about 200 to 300 miles inland of the Peru-Chile Trench. Yet the Foreland Ranges poke up 600 to 1,000 miles inland of North America's western edge (where the Farallon Plate used to subduct, before the development of the San Andreas fault; see figure 6.8). Even when we subtract the extra east–west distance added by the later stretching

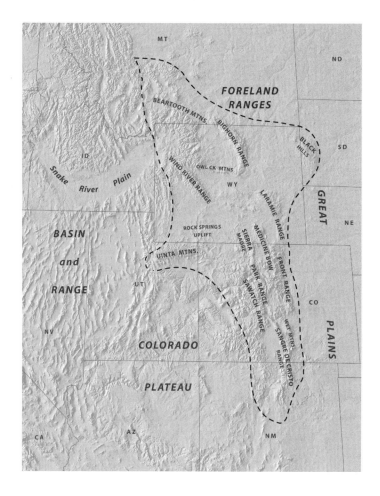

FIGURE 12.2

The Foreland Ranges dominate Wyoming and Colorado and extend into portions of Montana, South Dakota, Utah, and New Mexico.

of the Basin and Range—some 200 to 250 miles—it's clear that the Foreland Ranges occur anomalously deep in the continental interior.

Interior mountain belts exist elsewhere in the world. Many of them represent the glue-lines where once-separate continents have collided or continue to collide. Russia's Ural Mountains, for example, are the worn-down seam left from a stupendous continental train wreck about 300 million years ago—one of several continental collisions that assembled Asia. India's ongoing collision with southeastern Asia is presently hoisting mountains 1,400 miles into the continental interior (the Tian Shan mountains of western China). Saudi Arabia is plowing like a battering ram into the Middle East, crumpling up mountains 700 miles into the Asian interior (the Koppeh Dagh range of northern Iran and the Caucasus mountains of Georgia and Azerbaijan). But continental collisions don't help us explain the interior position of the Foreland Ranges. For more than a billion years, no continent-sized mass has crunched into western North America. More recently, late in Paleozoic time, a massive slab of seabed called the Golconda Terrane landed against North America's then-western edge, adding to the continent most of what is now western Nevada. More terranes crunched in behind Golconda during Triassic, Jurassic, and Cretaceous time, including the rocks that assembled California—those great bands of serpentinite, gabbro, and pillow basalt that we visited in chapters 1 and 2. As these terranes plowed into the continent, mountains sprouted in an east-migrating wave across Nevada and Utah. (These were fold-and-thrust mountains created by sideways compression, not related to the later stretching-induced mountains of the Basin and Range.) It's tempting to think that sideways pressure from terranes colliding against the west coast helped to crumple up these interior mountains. But whether it did or not, terrane collisions don't help us explain the Foreland Ranges. Those mountains started rising eighty million years ago, late in Cretaceous time, long after most of the west coast terranes had landed.

Bottom line: it's hard to explain the deep interior location of the Foreland Ranges either by normal subduction (which raises mountains mainly near the edges of continents), by continental collision (which hasn't happened in western North America for a long time), or by sideways pressure from terranes docking against the west coast. The Foreland Ranges demand a different explanation. To find it, we need to visit the ranges themselves and then head south to Argentina.

· · ·

Imagine sitting on a porch with a large hippo asleep underneath. As the hippo wakes and slowly stands up, his back pushes up through the splintering floorboards, stopping several feet above the level of the porch. The floorboards, once flat, now lean against the hippo's flanks. In miniature, the hippo's back represents the deep, uplifted cores of the Foreland Ranges. The floorboards represent younger rock layers that once lay overhead but which now tilt against the flanks of the ranges. We call such mountains basement-cored uplifts, because their uplifted cores are made of basement rock: the deep, ancient crystalline rock that forms the foundation of the continents. The basement rocks of the Foreland Ranges are mostly granites, gneisses, schists, distorted lava flows, and mashed-up sediments of Archean and Proterozoic age—a span that encompasses three billion years and includes some of the oldest rocks on Earth.

Before getting on with the story of how the lowest rock in North America came to form some of its highest mountains, let me take you on a brief detour down to the basement. North America's basement rock testifies to the forces that assembled the continent over the sweep of Archean and Proterozoic time. Some three billion years ago, long before there was any life on land, back when the seas contained nothing more biologically interesting than slippery bacterial goo, North America was in an adolescent growth spurt. Barren bits of crust, slashed by wind and rain, were swept into collision as new-formed plates jostled about on the face of the hot, young Earth. One by one, these pieces of primeval North America-to-be—mostly volcanic islands, slabs of dispossessed ocean floor, and assorted continental fragments—jammed together to build the continent outward by serial accretion at its edges. It was continental assembly on a grand scale, and it took place mostly before about 900 million years ago. Between about 900 million and 500 million years ago, erosion gradually beveled this crustal amalgam down to a low surface to become the basement—the stage upon which the rest of the continent's history has played out. Continental growth didn't end with the formation of the basement; North America continued to gain mass by accretion at its edges, particularly along the west coast (see figure 1.4). But these later accretions, although central to the geologic story of California and the rest of the west coast, represent minor additions to North America's total crustal inventory—some extra fat added to continental girth gained mostly during Archean and Proterozoic time.

In the United States and Mexico, basement rock lies mostly deep out of sight; we glimpse it only in the uplifted cores of mountains or in the bottoms of very deep canyons like the Grand Canyon. But in Canada, continental ice sheets have scraped away much of the younger rock that normally covers the basement. Consequently, basement rock sprawls in naked glory across two-thirds of Canada, defining a vast region known as the Canadian Shield. The oldest basement rocks yet discovered come from the Canadian Shield: 4.28-billion-year-old greenstone (metamorphosed basalt) from the Nuvvuagittuq Belt on the eastern shore of Hudson Bay. Given that the Earth is probably no older than about 4.56 billion years, it's not likely that we'll ever find basement rocks much older than those from the Nuvvuagittuq Belt.

Back now to the Foreland Ranges. Normally we expect to find basement rock down in the basement, with younger rock on top. But in the Foreland Ranges, the lowest, oldest rock of the continent pokes at the bellies of clouds. As I write this, a piece of sugar-textured quartzite sparkles from my bookshelf. One billion years old, it hails from a high, balding summit in Utah's Uinta Mountains. For most of its existence, it lay in the basement under five miles of younger rock. The rise of the Uinta Mountains removed that thick overburden, exposing the quartzite to ice, lichen, and, eventually, me. Beside it on my shelf sits a chunk of 1.42-billion-year-old Sherman Granite, lifted from its basement home four miles underground to become part of Wyoming's Laramie Range. Looking at it, I'm reminded of climbing 14,255-foot Longs Peak in Colorado's Front Range some years ago, where I splayed out—gasping and oxygen-deprived—onto comparably ancient basement granite. Oil wells near Denver have drilled into this same Longs Peak granite 13,000 feet below the city, proving that the Front Range has been jacked up more than four miles from its basement origins. My most cherished basement memento is a tiny flask of water, sealed with a cork stopper and labeled "Source of the Green River." It comes from an icy creek high on the ragged south face of the Wind River Range. To get it, my brother, Malcolm, hiked high above tree line in the Wind River Range to gather snowmelt dribbling from gneiss and granite as old as 3.1 billion years. That flask has double significance for me: first, for the water, with its connotations of

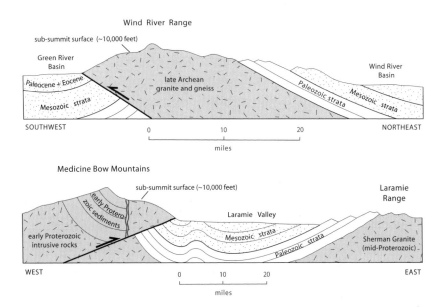

FIGURE 12.3

The Foreland Ranges are basement-cored uplifts—mountains formed from the up-squeezing of great blocks of basement rock. In some of the ranges, the basement blocks have squeezed upward as much as eight vertical miles along large thrust faults. In the diagram, basement rock is shaded gray, and younger sedimentary layers are shown with white or stippled patterns. The sub-summit surfaces notched onto the uplifted basement cores of the ranges are thought to record the maximum burial level of the mountains prior to the Exhumation (see figure 11.7).

John Wesley Powell and his adventures on the Green River, and second, for the basement rock that gathered that water, as snow, from the Wyoming sky—rock two-thirds as old as the Earth itself; rock dating to the very birth of our continent; rock normally found five or more miles underground. Throughout the Foreland Ranges, the oldest, deepest rock of the continent has risen into thin air, and younger rock lies far below, leaning in great hogbacks against the uplifted basement cores. The prosaic term "basement-cored uplift," when you think about what it means, encapsulates a mind-bending flip in the normal order of things.

To understand the Foreland Ranges, we need to figure out the mechanism that pushed them up. The ranges are essentially large wedges of basement rock squeezed upward by colossal sideways pressures. They rose along ramp-like faults called thrust faults in which the rock above the fault was thrust up and over the rock below (figure 12.3). The ranges began to jolt upward about eighty million years ago and continued to do so, earthquake by earthquake, until about forty-five million years ago. (We infer the timing from the age of sedimentary debris shed from the rising ranges into their adjacent basins.) By the time they had stopped, Colorado and Wyoming had been squeezed sideways to roughly four-fifths of their original width (figure 12.4).

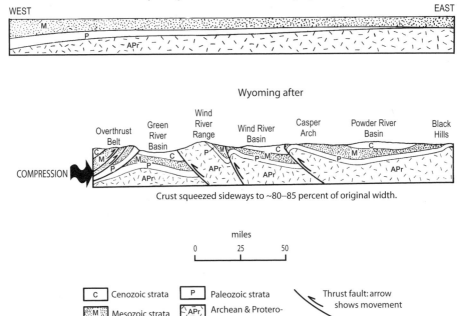

FIGURE 12.4

The Foreland Ranges rose in response to massive sideways squeezing during the Laramide Orogeny from eighty million to forty-five million years ago. The diagram illustrates the changes that the orogeny wrought in Wyoming. Compression from the west-southwest squeezed the crust like an accordion, wedging large blocks of basement rock upward to form the ranges. Areas in between the ranges sagged down to become basins that filled with rock debris shed from the ranges. (Reproduced with permission of the Wyoming State Geological Survey.)

Looking back to figure 12.2, you can see that the Foreland Ranges generally trend north to south or northwest to southeast. (The Uinta and Owl Creek mountains buck the trend by running east to west for reasons that remain elusive, although it may reflect the reactivation of much older faults.) The grain of the ranges reflects the alignment of the faults that lifted them; like the ranges, the faults line up mostly north to south or northwest to southeast. This tells us that the sideways-squeezing forces that pushed up the ranges came mostly from the west-southwest, perpendicular to the linear trend of the mountains. Somehow, between eighty million and forty-five million years ago, massive sideways pressure compressed the continental basement below Colorado and Wyoming from a west-southwest direction. But how? As I argued earlier, no conventional mechanism of mountain uplift will get the job done.

Geologists are always happier about explanations for things in the past if they can see the Earth doing similar things today. To understand how large blocks of basement rock began to squeeze up in the deep interior of a continent some eighty million years ago, we can first ask whether similar mountains are squeezing up in a continental interior somewhere on Earth today.

As it turns out, they are.

In the Pampas region of Argentina, several hundred miles inland from the main chain of the Andes, modern versions of the Foreland Ranges are squeezing out of the basement right now. Although the Pampean Ranges don't quite match the Foreland Ranges in scale and grandeur (not yet anyway—they're young and still growing), their basic structure is identical. Like the Foreland Ranges, the Pampean Ranges are basement-cored uplifts jolting upward along huge thrust faults as they escape sideways pressure from the west. And like the Foreland Ranges, they are rising deep within the continental interior—as much as 300 miles east of the Andes and 600 miles east of where the Nazca Plate plunges into the Peru-Chile Trench. They began to poke up about eight million years ago, and they are still going—as the earthquakes that regularly rattle this region of the Pampas attest. The largest of the Pampean Ranges, the Sierra de Córdoba, is like a twin of Colorado's Front Range. It trends north to south, occupies a frontal (easternmost) position among its brethren, casts its afternoon shadows across great plains, and even has the largest city in the region, Córdoba, nestled at its foot like a geologic sister city to Denver at the foot of the Front Range.

If Argentina's Pampean Ranges represent modern, still-forming versions of the Foreland Ranges, what's pushing them up? Answers have come from tracking seismic waves through the Earth's interior. This technique, called seismic tomography, allows us to take grainy pictures of the Earth's innards—somewhat like ultrasound images—by tracking earthquake waves as they radiate through the planet. Remarkably, our seismic view under the Pampean Ranges shows that the Nazca Plate subducts *nearly flat* beneath South America here, like a board sliding under a carpet. Nearly everywhere else, the Nazca Plate subducts at a steep angle beneath South America, yielding the archetypal subduction-generated mountain belt of the Andes, complete with active volcanoes (figure 12.5a). But in a 350-mile-wide zone between about 28 degrees and 33 degrees south latitude, the Nazca Plate appears to flatten against the bottom of South America (figure 12.5b). Consequently, the once-vigorous volcanoes here have sputtered out. As the Nazca Plate flattened, some fifteen to ten million years ago, it evidently rose above its magma-generation depth, shutting off the supply of molten rock to the volcanoes. By about eight million years ago, the flattening plate seems to have bumped into the bottom of South America like a submarine trying to surface through pack ice. It began scraping eastward underneath the continent, compressing the continental basement from the west—and the Pampean Ranges began to wedge upward in response (figure 12.5c).

The events that I've just described in Argentina echo those that occurred during the Laramide Orogeny in the western United States. From about eighty million to forty-five million years ago, the Farallon Plate probably subducted flat in the 500-mile-wide zone

Normal subduction: occurring under most of the Andes Mountains

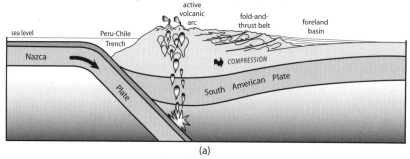

(a)

Flat subduction: occurring between ~28° and 33° south latitude

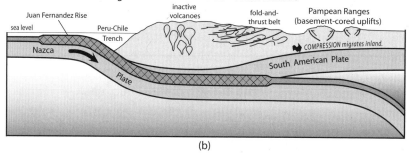

(b)

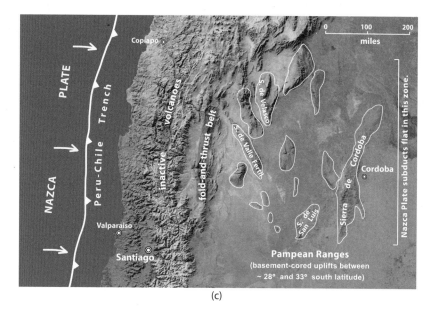

(c)

FIGURE 12.5

The Pampean Ranges—modern versions of the Foreland Ranges—are rising today in the Pampas region of Argentina several hundred miles east of the Andes. Diagram (a) illustrates the situation under most of the Andes today: classic subduction-generated volcanism and mountain building. Diagrams (b) and (c) illustrate the situation where the Pampean Ranges are forming. Like the Foreland Ranges of the western United States, the Pampean Ranges are basement-cored uplifts that pop up far inland of the main mountain belt. The reason seems to be the nearly flat subduction of the Nazca Plate in this area.

between southern Montana and northern New Mexico. In response, the Foreland Ranges wedged out of the basement, the Colorado Plateau buckled and warped, and once-active volcanoes to the west in Nevada and Utah sputtered out. All three events make sense with flat subduction. Nothing comparable occurred north into Canada or south into Mexico. In these areas, the Farallon Plate seems to have subducted at a more or less normal angle during the Laramide Orogeny, producing the signature features of Andean-style mountain building: abundant volcanism and the rise of mountains closer to the continent's edge. Starting about forty-five million years ago, the flat portion of the Farallon Plate seems to have bent back down to a normal subduction angle. The Foreland Ranges stopped rising, the Colorado Plateau stopped buckling, and volcanoes reawakened with a vengeance across Nevada and Utah, where the Great Basin would later form. If you flip back to figure 8.3, you can see this sequence of events portrayed in tectonic cross sections.

As you may remember from chapter 8, the volcanism that came in the aftermath of the flat subduction episode was stunning in its scope and violence. From about forty-three to twenty-one million years ago, calderas opened fire across the future Great Basin, carpeting the landscape with layer after layer of scalding lava and ash. It was volcanic mayhem on an unprecedented scale—and flat subduction during the Laramide Orogeny may be to blame. As the Farallon Plate slid flat under the continent, the surrounding pressures may have squeezed quantities of old seawater out of the plate. The water might then have percolated up into the lower reaches of the continental crust. When the Farallon Plate angled back down, the inflowing soft mantle above the plate began to melt. When this fresh magma rose up and entered the now-hydrated crust above, all hell broke loose. For one thing, the presence of water enhances magma formation by lowering the melting point of rock. For another, water-saturated magma is one the Earth's most violent forces. When such magma rises toward the surface and depressurizes, the water converts to steam and the magma can explode with the force of multiple hydrogen bombs, detonating calderas as much as fifty miles wide. We see this violent legacy written in the volcanic rocks of the Great Basin. The history of the Farallon Plate suggests that when the Nazca Plate eventually unflattens under the Pampean Ranges, volcanic Armageddon will come to Chile and Argentina. If people are still around when that happens, they had better clear out—because anyone who stays behind to watch the show will end up looking like Argentine beef abandoned on the barbeque grill.

. . .

If you accept the evidence that flat subduction made the Foreland Ranges, just as it appears to be making the Pampean Ranges today, you're probably wondering how and why flat subduction can happen. To find out, we need to look at the processes that control subduction.

When an oceanic plate first forms by seafloor spreading, it is hot and buoyant. But once it has aged and cooled for about ten million years, it becomes denser than the hot, mushy mantle beneath it and thus wants to sink. Where oceanic plates lie flat on the mantle, they won't sink because the viscous resistance of the mantle is too great.[3] But once the edge of an oceanic plate begins to bend down to form an ocean trench, the leading edge of the plate will naturally sink, pulling the rest of the plate along behind it much in the way that a towel hanging half off a table pulls itself to the floor. This process, called slab pull, is the main driving force behind the movement of the Earth's plates. Many cartoon views of the Earth show the mantle churning furiously via convection, and such images give the impression that mantle convection is what pushes the plates around. But this isn't correct. Plates move mostly because they are pulled at their subducting edges, not because the churning mantle pushes them. We know this, in part, because plates with longer subducting edges move faster. The Pacific Plate is the world's fastest plate (sliding northwest at more than three inches per year), and the reason seems to be that it is tethered to the longest system of ocean trenches down which some of the oldest—and thus coolest and densest—ocean floor on Earth is plunging. Plates with shorter subducting edges (the Nazca Plate or the Australian Plate, for example) are pulled more slowly toward their respective trenches—but still, they are pulled. Plates with minor or no subducting edges simply don't move much.

The key point is this: most of the time, oceanic plates sink down ocean trenches because they are denser than the mantle beneath. But if part of a subducting oceanic plate becomes *less* dense than the mantle, it will become too buoyant to sink. The rest of the plate on either side may continue to sink; the buoyant part isn't going to change that any more than a lifejacket would have slowed the sinking of the *Bismark*. But the buoyant section, instead of angling downward with the rest of the plate, will float upward in a great arch. As the rest of the plate continues to subduct in normal fashion, that upward-floating arch will slide flat beneath the plate overhead and scrape against it. In theory, the arch may even tear away from the rest of the plate like a flap torn along the edge of a piece of paper.

To get flat subduction to happen, therefore, we need to figure out how one portion of an oceanic plate can become less dense than the mantle—so that it floats upward and subducts flat—while the rest of the plate stays at normal density and thus continues to subduct in normal fashion. This can happen when events alter the proportions of the different types of rock that make up oceanic plates. Viewed simply, oceanic plates are like two-ply plywood, consisting of a thin top layer of lightweight basalt and gabbro (basalt's slow-cooling equivalent) and, below that, a much thicker layer of very dense peridotite. Most of the time, the weight of that high-density peridotite overcomes the relatively lightweight rock above it, so that the net effect is for the plate to be quite dense and to sink. But there's a way to change that. An oceanic plate can become buoyant in the mantle if a lot of extra basalt and gabbro is added to its top. By analogy, visualize a waterlogged dock that has sunk to the bottom of a lake. You dive down and nail layers of fresh, buoyant wood to the top of the dock. Hammer in enough buoyant wood, and the dock will rise to float again. Density calculations suggest that if we add about ten extra miles of basalt to the top of an oceanic plate, the plate overall will become less dense than the mantle and no longer sink. Instead, that extra layering of low-density rock will cause the plate to subduct flat, or nearly flat, beneath the plate overhead.

Such a buoyant thickening appears to explain the flat subduction of the Nazca Plate today beneath Chile and Argentina. At about 32 degrees south latitude, a thick basalt ridge called the Juan Fernández Rise sits on the Nazca Plate. Seismic images suggest that much of the Juan Fernández Rise has already subducted down the Peru-Chile Trench (figure 12.5b). The extra buoyancy that the Juan Fernández Rise gives the Nazca Plate is probably causing the flat subduction under this part of Argentina.

Did a similar thickening of basalt on top of the Farallon Plate cause it to flatten under North America, thus pushing up the Foreland Ranges? If we look around the ocean floor today, we find several regions of extra-thick seabed that—if they were subducted— might cause such a flattening. These areas, called oceanic plateaus, form where vast quantities of basalt lava vent onto the abyssal seabed. These plateaus rise from the sea-floor like great pancakes of basalt, up to several hundred miles across and as much as twenty miles thick (four to five times thicker than average for the ocean floor). The world's largest oceanic plateau is the Ontong-Java Plateau east of New Guinea. Two times the size of Texas and up to twenty miles thick, it contains enough basalt to cover the lower forty-eight U.S. states more than two miles deep. (In fact, the Ontong-Java Plateau is so thick that its top breaks the ocean surface to form the Ontong-Java Atoll: the world's largest coral atoll.) Shove the Ontong-Java Plateau underneath eastern Asia, and mountains like the Foreland Ranges might rise deep inside Siberia.

It wouldn't have taken something as big as the Ontong-Java Plateau to push up the Foreland Ranges. A ten-mile-thick basalt pancake about the size of Texas, subducted under the western United States on the Farallon Plate, would have gotten the job done.[4] Had that not happened the Foreland Ranges would likely not exist. The Great Plains would reach west to Utah, and Colorado and Wyoming would look like Kansas and Nebraska. Denver would be the Half-Mile-High City, at best, with mountain views as exhilarating as Wichita's.

At least, that's part of the story. The other part relates to what happened next, namely, the Exhumation—the digging out of the Rockies from their deep Miocene burial. The fact that the Foreland Ranges today soar thousands of feet above the Great Plains and the intermountain valleys owes as much to their unburial during the Exhumation as it does to the original uplift of the mountains (figure 12.6). Wherever mountains rise in the world, erosion soon counters. As the Foreland Ranges began to go up, wind, rain, and ice began to take them down. Streams swept the eroded debris of the mountains out into the neighboring valleys and across the Great Plains. Up and up the layers rose, like a gravel tide against the mountainsides, assisted by periodic blizzards of volcanic ash blown in from calderas erupting across the Great Basin and the Snake River Plain to the west. By late in Miocene time, about five to ten million years ago, the Foreland Ranges were so deeply buried that Colorado and Wyoming looked like somewhat lumpy

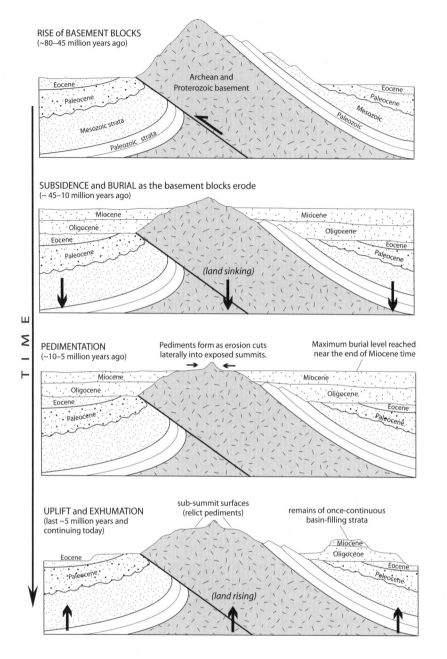

FIGURE 12.6

The history of the Rocky Mountain Foreland Ranges summed up in four time panels. We'll explore the causes of the Exhumation—the event portrayed in the bottom panel—in chapter 13.

versions of Iowa or Indiana. Then—for reasons that we'll explore in the next chapter—the once-lazy rivers of the region began to bite downward. They bulldozed away the debris that had covered the mountains, and brought the Rockies back to the world. Nowhere is the work of this monumental excavation on better display than where we are headed next—to the western Great Plains.

NOTES

1. To save you the trouble of running for your atlas, the top 20 U.S. states by average elevation are: Colorado 6,800 feet, Wyoming 6,700, Utah 6,100, New Mexico 5,700, Nevada 5,500, Idaho 5,000, Arizona 4,100, Montana 3,400, Oregon 3,300, Hawaii 3,000, California 2,900, Nebraska 2,600, South Dakota 2,200, Kansas 2,000, North Dakota 1,900, and Alaska 1,900.

2. Readers familiar with Rocky Mountain geography may wonder at three ranges missing from this list: the San Juan Mountains of southwestern Colorado, and the Absaroka Mountains and Teton Range of northwestern Wyoming. All are younger than the Foreland Ranges, and they formed in a different way. The Foreland Ranges formed by the upward squeezing of great blocks of deep basement rock between about eighty and forty-five million years ago. The younger Teton Range formed as part of the stretching of the Basin and Range Province during the last twenty million years. The San Juan and Absaroka mountains both formed by voluminous volcanism. The Absaroka volcanics erupted mostly from forty-nine to forty-four million years ago, just as the Foreland Ranges had finished jolting upward. The San Juan volcanics erupted mostly from thirty-two to twenty-six million years ago, coinciding with—and possibly related to—the cataclysmic caldera eruptions of the Great Basin that we explored in chapter 8.

3. Whether the mantle below the plates is best thought of as a solid or an extremely thick, slow-flowing liquid is a matter of perspective. An analogy might help here. Intuitively we might think of piano wire as solid, but in fact, it stretches slowly—a fact that keeps piano tuners in business. Piano wire over a long period of time behaves somewhat like a liquid because it changes its shape in response to an external force. In a sense, piano wire flows. The viscosity (meaning the resistance of a substance to flow) of the hot mantle below the plates is probably close to the viscosity of piano wire. This analogy is from McPhee 1998 (48).

4. See Liu et al. 2010 for a compelling argument that the oceanic plateau that was subducted to raise the Foreland Ranges has a conjugate on the Pacific seafloor today called the Shatsky Rise.

13

AT THE FRONTIER

Nearly everywhere from Montana to New Mexico, the Rocky Mountains face the Great Plains like a mighty wall. But if you approach the mountains along the route of Interstate 80 in southeastern Wyoming, they appear much less formidable. From western Nebraska, the interstate rises like a smooth ramp all the way to the summit of the Laramie Range. It's as if someone had laid a great plank across the landscape, with one end on the Great Plains and the other on the roofline of the mountain. When, in 1865, surveyors for the Union Pacific Railroad found this smooth route from the plains to the top of the Laramie Range, they were delighted. No need for costly tunnels or switchbacks here. They hammered their survey stakes into the ramp-like surface and launched the transcontinental rails across the Rockies.

The Cheyenne Tableland, as this region is known (figure 13.1), is a piece of fossil geography. It's a relic of the Great Plains as they appeared about five million years ago, near the end of Miocene time. At that time, the sand and gravel layers of the Great Plains sloped so high against the eastern face of the Rockies that they covered the lowest summit ridges and brushed the chins of the higher peaks. Back then, you could have walked west across the Great Plains and stepped off directly onto the high flanks or even the summits of the Rocky Mountains. That's how we would see the eastern face of the Rockies today—buried, or nearly so, under thick blankets of sand, gravel, and volcanic ash—if not for the Exhumation. During the last five million years, rivers have scoured the former plains away from the mountains nearly everywhere but on the Cheyenne Tableland. Looking at figure 13.1, restore in your mind all the western Great

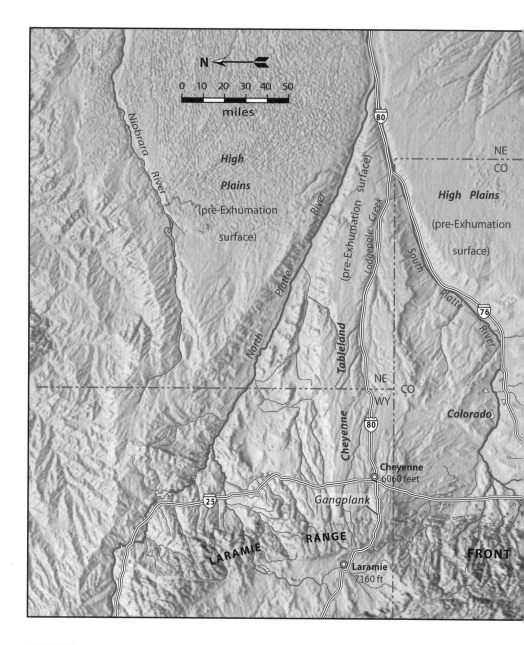

FIGURE 13.1

The effects of the Exhumation on the western Great Plains show up boldly on this shaded relief image. The map is oriented so that north is to the left. Notice the smooth, east-sloping surfaces labeled High Plains in the eastern (upper) areas of the map. These represent surviving remnants of the Miocene Great Plains, which once reached unbroken all the way to the Rocky Mountains. To the west, nearer the Rockies, rivers such as the North Platte, South Platte, and Arkansas have bitten deeply into these former plains to carve out broad valleys, including the Colorado Piedmont.

Only on the Cheyenne Tableland does the original Miocene Great Plains surface still reach all the way to the Rockies. Gravels of the Ogallala Formation, washed east from the Rockies by Miocene streams, cap both the High Plains and the Cheyenne Tableland. Notice the location of the Gangplank (along Interstate 80 west of Cheyenne), and of Mount Evans (south of Interstate 70 where it passes through the Front Range), both of which we visit in this chapter.

Plains to the level of the Cheyenne Tableland. (In other words, fill in those big, river-carved valleys centered on the North Platte, South Platte, and Arkansas rivers.) You look then onto a Miocene world, with the plains at their maximum advance against the mountains.

In the 1950s, President Dwight Eisenhower dispatched surveyors across the nation to lay out the grid of the new national interstate highway system. The surveys were thorough, with multiple routes proposed and debated. But in one place, there was no debate. Interstate 80 would climb the Rocky Mountains using the Cheyenne Tableland. There was no better way across the Wyoming Rockies.

Interstate 80 first gains the tableland in western Nebraska, after climbing out of the cottonwood-clad valley of the South Platte River. As you drive the westbound lanes, the flow of Lodgepole Creek (which the interstate closely follows) tells you that you must be going uphill. But you're hard-pressed to detect any gradient. Over the next 160 miles, you climb all of three-quarters of a mile—a slope equal to that of a ten-foot-long table lifted a half-inch at one end. All around, you see wheat, wheat, wheat. But just a century-and-a-half ago, the Cheyenne Tableland was wild prairie where nomadic tribes of Cheyenne and Sioux chased after wandering herds of buffalo. In 1867, as crews for the Union Pacific Railroad spiked rails west across the tableland, the Indians attacked them with vigor. Yet nothing would stop the railroad; it was the ultimate symbol of national unity and progress in the wake of the Civil War. The U.S. Army, having finally crushed the Confederacy, headed west to deal with the Indians. General John Pope summed up the conundrum of the task. "The Indian's first demand is that the white man shall not drive off his game and dispossess him of his lands. [Yet] how can we promise this unless we prohibit emigration and settlement? . . . The end is sure and dreadful to contemplate." Few other army commanders were so circumspect. "We've got to clean the damn Indians out or give up building the Union Pacific Railroad," General Grenville Dodge groused, adding, "Until they are exterminated, or so far reduced in numbers as to make their power contemptible, no safety will be found."[1] The greatest advocate of Indian extermination was Union war hero William Tecumseh Sherman. His middle name, picked by his father to honor a Shawnee chief, had no discernable effect on Sherman's feelings about Indians. "The more we can kill this year, the less will have to be killed the next year," he asserted, "for the more I see of these Indians the more convinced I am that they will have to be killed or maintained as a series of paupers." Without embarrassment, Sherman promoted a policy of genocide. "We must act with vindictive earnestness against the Sioux, even to their extermination, men, women, and children," he wrote. In the years that followed, that's more or less how it went.

The great ramp of the Cheyenne Tableland takes you smoothly out of Nebraska into Wyoming, through the city of Cheyenne, and beyond, up more than 7,000 feet above the sea. You see no wheat fields now. Even in summer, the temperatures here can drop below freezing. On both sides of the tableland, the excavation of the former Great Plains has been deep and profound, as if two hungry men had eaten a layer cake from

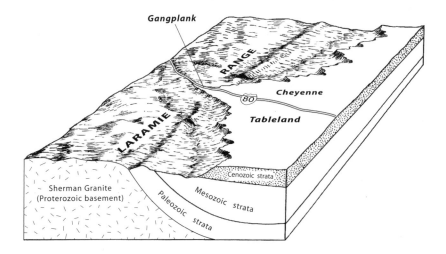

Gangplank

RANGE

Cheyenne
Tableland

LARAMIE

80

Cenozoic strata

Sherman Granite
(Proterozoic basement)

Paleozoic strata

Mesozoic strata

FIGURE 13.2

The Cheyenne Tableland, as it rises west toward the Laramie Range, narrows to a finger-like ridge called the Gangplank. Here the gravelly layers of the upper Miocene Ogallala Formation lap smoothly onto the mountain's uplifted core of Proterozoic granite. The Cheyenne Tableland / Gangplank forms a smoothly ascending ramp up onto the Laramie Range. The transcontinental railroad of 1867 took advantage of this route to cross the Laramie Range, and Interstate 80 follows the same route today. (Reproduced with permission of the Wyoming State Geological Survey.)

opposite sides, leaving only a narrow plateau in the middle, now topped by the interstate. To the south, the Front Range looms grandly over the exhumed plains. To the north, the northern Laramie Range does the same. But dead ahead, along the rising tableland surface, you face no mountain escarpment. It's as if you're driving a great ramp up into the Wyoming sky.

The ramp gradually narrows, and soon there is no room to spare. You are now on the Gangplank—the skinny, westernmost finger of the Cheyenne Tableland (figure 13.2). It is a tenuous ridge, a narrow bridge of relict Miocene gravel that carries you the final miles up onto the Laramie Range. The interstate and the railroad tracks squeeze close and run in parallel. The heavy freight cars clank to your right; the traffic of the eastbound interstate lanes blurs by on your left. Down below, off the gully-etched flanks of the Gangplank, you see tilted Paleozoic layers leaning west like a row of books tipped over on a shelf. You know what's coming next—the basement core of the Laramie Range, which, as it rose from below, pushed those layers aside. Smoothly, the Gangplank launches you from Miocene gravel (five-million-year-old Ogallala Formation, the capping stratum of the former Great Plains) onto Proterozoic granite (1.4-billion-year-old Sherman Granite, the core rock of the Laramie Range). Some miles ahead, the highway passes over the bald granite crest of the Laramie Range at Sherman Summit, 8,640 feet above the sea—the highest point on the cross-country route of Interstate 80. Although 1,550 feet higher than Donner Pass in the Sierra Nevada, Sherman Summit is

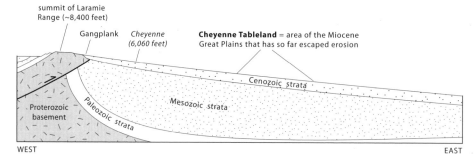

Route of Interstate 80 from the Great Plains to the Laramie Range via Cheyenne, Wyoming

Route of Interstate 70 from the Great Plains to the Front Range via Denver, Colorado

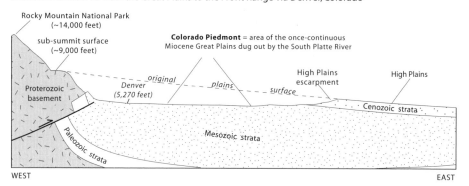

FIGURE 13.3

Contrasting approaches to the Rocky Mountains from the Great Plains. Interstate 80 (top) follows the Cheyenne Tableland—a remnant of the Miocene Great Plains—to ascend nearly to the summit of the Laramie Range. Interstate 70 (bottom), after climbing the same remnant Miocene plains (the High Plains of eastern Colorado) for some distance, drops off the High Plains Escarpment into the Colorado Piedmont, a great bowl carved out of the former plains by the South Plate River system during the last five million years. A prominent sub-summit surface on the eastern face of the Front Range marks the level of the former plains.

much easier to reach by road and rail, thanks to the Cheyenne Tableland and the Gangplank—surviving relics of the Miocene Great Plains.

If you approach the Rockies along Interstate 70 or 76 across Colorado, your experience is entirely different, as you can see by comparing the two illustrations in figure 13.3. Whereas I-80 passes smoothly *over* the Laramie Range, I-70 passes torturously *through* the Front Range along a winding route framed by high mountain walls. The difference relates, in part, to the unevenness of the Exhumation on the western Great Plains.

Interstate 70 climbs the High Plains for several hundred miles across Kansas into Colorado. (The High Plains, marked on figure 13.1, represent the same relict Miocene surface as the Cheyenne Tableland, capped by the same gravelly beds of the upper Miocene Ogallala Formation.) As you approach the town of Genoa, about eighty miles east of Denver, the Front Range becomes visible ahead in clear weather. Project the rising slope of the High Plains west from here, and you would seem to be on track to be launched into the Front Range at an elevation of about 9,000 feet. But that's not what happens. Instead, the interstate drops off the High Plains Escarpment into the Colorado Piedmont: a vast, riverine bowl dug out of the former plains by the South Platte River and its tributaries (figure 13.1). As you cross the bowl-like piedmont, the Front Range looms higher. Eventually, you spot a broad, bench-like surface about halfway up the mountainside, looking like a rough step cut into the eastern face of the Front Range. If you project the slope of the High Plains westward, you intersect this step-like surface, which was once continuous with the former plains (figure 13.3). Local geologists know it as the Rocky Mountain sub-summit surface or the Tertiary pediment. Restore the former plains to this level, and Denver would be under more than a half mile of gravel.[2]

The interstate arrows through Denver and plunges into the Front Range through the Dakota Hogback—a spectacular ridge of Mesozoic layers upended by the rising basement core of the range. The highway begins winding uphill through steep-walled canyons that glitter with dark schist, 1.8 billion years old. This basement rock speaks of a time when Colorado looked much like Indonesia: a collection of scattered archipelagoes divided by areas of deep ocean floor. The schist is the squeezed and distorted remains of sand, mud, and lava flows that sloped off the flanks of these islands into the sea. Eventually, as the trenches between the islands closed, the islands bunched together to assemble this part of the North American continent.

After winding and climbing for a time, the interstate intersects the Mount Evans highway—a switch-backing, carsickness-inducing ribbon of potholed pavement that takes you uphill to within a few steps of the 14,264-foot summit of Mount Evans (marked on figure 13.1). As befitting the nation's highest state, Colorado's highway to Mount Evans is the nation's highest paved road. It's also a cheater's way to bag a Colorado fourteener, but who cares? The view is great, and more to the point, there are few better perches from which to take in the work of the Exhumation on the western Great Plains.

From the summit of Mount Evans, half of Colorado spreads before you to the east. (In clear weather anyway; the best views come after rainstorms have swept the air clear of dust and smog.) Although only thirty crow-miles from Denver, the snow-patched tundra landscape all around looks more like Baffin Island. The barren granite slopes plummet to tree line more than 3,000 feet below. There, in the trees, at roughly 8,000 to 9,000 feet elevation, the landscape flattens into an undulating, forested plateau bisected by steep-walled stream valleys. That plateau is the Rocky Mountain sub-summit surface—the same one that you looked *up* at from the Colorado Piedmont east of Denver. Five million years ago, this sub-summit surface merged seamlessly with the surface of the Great Plains. Now, it sails off into open space above a cavernous bowl filled with Denver smog. To excavate that bowl—the Colorado Piedmont—the South Platte River and its tributaries had to dig out some 2,000 cubic miles of gravel during the last five million years. That's considerably *more* than the volume of the Grand Canyon, which the Colorado River cut during roughly the same time span. In other words, more than three million Coloradans live in a river-carved hole larger than the famous one in Arizona, cut during the same span of time.[3] And that's just the work of the South Platte River system. Looking at figure 13.1, you can see that the Arkansas River to the south and the North Platte River to the north have both carried out their own monumental excavations of the western Great Plains. Turning brown with each rainstorm, the rivers of the Great Plains annually ship tons of sediment eastward to the Mississippi River (or, more exactly, into the intervening reservoirs, which, in a geologic eye blink, will overtop with silt). The Exhumation of the Rocky Mountains is a work in progress.

·   ·   ·

In the last two chapters, we've witnessed the work of the Exhumation throughout the Rocky Mountains. We've seen it in the paths of the range-roving rivers. We've seen it in the high sub-summit surfaces that mark the former burial depth of the mountains like a high-tide line. We've seen it in the dissected layer cakes of strata in the intermountain valleys. And we've seen it in the riverine excavation of the western Great Plains.

But why did the Exhumation happen? Why don't we see the Rockies today as they appeared in Miocene time—deeply buried, with the Great Plains banked high against their eastern faces?

There are two competing theories.

The first is regional uplift. According to this theory, beginning about five million years ago,[4] the Rocky Mountain region began slowly rising. As the region rose, rivers flowed faster in response to the steeper slopes. The invigorated rivers sliced away the sedimentary layers that once covered the Rockies, and bit deeply into the former Great Plains.

The second theory is climate change. This theory holds that, starting about five million years ago, more intense rainstorms and deeper snow packs in the Rockies produced bigger floods. Since rivers do most of their erosive work during floods, the results would be similar to those expected from uplift: the exhumation of the Rockies, and the riverine excavation of the Great Plains.

If the uplift theory is right, then the Rocky Mountain region must have been several thousand feet lower before the Exhumation began. If the region was not lower, then uplift cannot explain the Exhumation, leaving climate change as the more likely explanation. The question thus hinges on knowing the elevation of the Rockies before the Exhumation.

The problem is, there isn't any foolproof way to tell the elevation of the pre-Exhumation Rocky Mountains. Researchers have tackled the problem from various angles, with conflicting results. Those who study fossil plants and rock chemistry can't find much evidence for uplift, and so they favor the climate theory. Those who study the rock layers of the Great Plains and how much they tilt (more about this below) favor the uplift theory.

My explorations of the controversy lead me to think that both the uplift and climate theories are partly right and that the Exhumation comes down to three factors that have all worked toward the same end during the last five million years: (1) passive uplift caused by erosion, (2) active uplift caused by a hot, buoyant mantle beneath the Rockies, and (3) climate change that delivered bigger floods to Rocky Mountain rivers. Let me explain all three, in sequence.

First, passive uplift. To some degree, the Exhumation has probably been self-enhancing. As rivers have excavated the Rocky Mountains, the region has bobbed upward, steepening the riverbeds, which in turn has prompted further excavation. To see how this can happen, we need to understand that the Earth's crust floats buoyantly in the denser rock of the mantle beneath. Remove weight from a floating object, and it will float higher. Icebergs serve as a useful analogy. Visualize cutting a deep groove, like a valley, through the middle of a large iceberg. The berg pops up in response to the weight loss, lifting the uncut parts of the berg higher above the water than they were before. Likewise, when erosion removes rock from mountain valleys, the intervening peaks rise higher. We see this happening in the Swiss Alps today. The Alps were shoved up when the Adriatic Plate crunched into the Eurasian Plate during Oligocene and Miocene time. Those plates have now mostly jammed to a halt, yet GPS measurements show that many Alpine peaks are still rising by a fraction of an inch each year. The reason is erosion. As glaciers bulldoze rock from Alpine valleys, the lightened mountains are bobbing upward in response. The Rockies seem to be doing the same thing—bobbing upward as rivers carry debris off the mountains.

All well and good, but passive uplift, by itself, could not have *caused* Exhumation; it could only *reinforce* it. There still needed to be a driving force—something to make Rocky Mountain rivers dig in and keep digging. The options are active uplift (something pushing the Rocky Mountain region up) or climate change that triggered larger and more erosive floods along Rocky Mountain rivers.

A hot, buoyant mantle beneath the Rockies could have provided the boost needed to raise the region and cause the Exhumation. Geologists have pondered this notion for a long time, but until recently they couldn't easily test the idea. That has now changed. Today, we can peer into the mantle beneath the Rockies with unprecedented clarity— thanks to USArray.

USArray is a grid of mobile seismometers—sensitive instruments that detect subtle ground vibrations from distant earthquakes. As seismic waves travel through the Earth, they speed up, slow down, and ricochet off in new directions as they encounter changes in the rock. By measuring the speed and direction of seismic waves coming from many earthquakes in different places, we can build up a picture of the Earth's insides, sort of like an ultrasound image of the planet. This technique—called seismic tomography— hinges on having many seismometers deployed and leaving them in their sheltered underground bunkers long enough to detect a lot of seismic waves. This is where US-Array is exceptional. USArray consists of a rectangular grid of 400 movable seismometers spread across a wide swath between the Canadian and Mexican borders. The first seismometers were installed along the west coast in 2004. Since then, the grid has expanded east. Once a seismometer on the west side of the grid has been in the ground for about two years, technicians dig it up and leapfrog it to the east side of the grid. In this manner, the entire grid of 400 instruments is slowly rolling east, like an MRI scanner sweeping across the nation. As I write this, the USArray grid has reached the Gulf Coast, and when you read it, the grid may well have reached the eastern seaboard.

USArray has given us some of the clearest views ever of the mantle beneath North America. Figure 13.4 shows one result: a map of mantle temperatures 120 miles under the western United States. Notice that the mantle is exceptionally hot under most of the Basin and Range (centered on Nevada and extending into parts of adjacent states). This comes as no surprise. Heat flows out of the ground at exceptional rates throughout much of the Basin and Range, and the region lies at unusually high elevations given the thinness of its stretched crust. A giant dome of hot mantle, welling up from below, explains these features. But for our purpose here—finding a force of uplift that might have caused the Exhumation of the Rocky Mountains—what's most interesting is that pocket of hot mantle that you can see under Colorado in figure 13.4. That, in all probability, is our smoking gun: the buoyant force that lifted the Rockies and caused the rivers to dig the mountains out of their gravelly grave.

But hold on. If we're going to believe that that hot mantle under the Rockies caused the Exhumation, we had better find some direct evidence that significant uplift has actually occurred—more uplift than just the passive up-bobbing that we expect to see anyway from erosion. To find that evidence, we need to go back to the Cheyenne Tableland—that surviving remnant of the Miocene Great Plains that still rises, unbroken, nearly to the top of the Laramie Range.

Geologist Margaret McMillan and colleagues have figured out a clever way to use the Cheyenne Tableland to measure how much the Rocky Mountain region has risen since the Exhumation. Here's the reasoning: If the Rockies have been uplifted since the sedimentary layers that form the Cheyenne Tableland were laid down, then those layers today will tilt east more steeply than when they formed. Picture a balloon under one end of a long board. Pump air into the balloon (the rising Rockies), and the board (the Cheyenne Tableland) rises on one end and tilts more steeply to the east. Today, the Cheyenne Tableland tilts east at slopes that increase from about ten feet per mile on its east end to fifty feet per mile on its west end, near the Gangplank. How does this compare to the slope of the tableland before the Exhumation? It turns out that the capping stratum of the tableland—the upper Miocene Ogallala Formation, which was laid down just before the Exhumation kicked in—contains sand and gravel deposited by braided streams.[5] This is fortuitous, because the maximum size of gravel that braided streams can move depends not on water volume but rather the slope of the streambed. By measuring gravel sizes in the Ogallala Formation, McMillan and colleagues were able to figure out the original Miocene slopes down which these braided steams flowed. They found that during Miocene time the Cheyenne Tableland sloped east at only a half

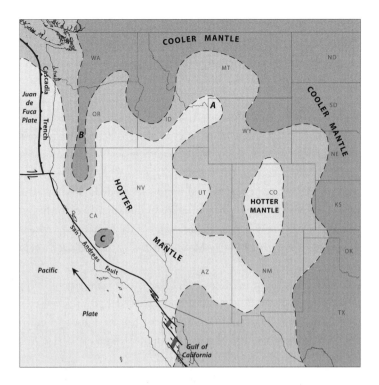

FIGURE 13.4

A map of mantle temperatures 120 miles underground, based on seismic wave speeds measured by the USArray seismic grid. Seismic waves move faster through cool rock and slower through hot rock. Notice that hotter mantle underlies all of Nevada and much of California, Utah, and Arizona, corresponding to the warm, stretching crust of the Basin and Range. The area of hotter mantle directly under Colorado may be responsible for the regional uplift of the Rockies that triggered the Exhumation. Label A marks Yellowstone National Park, located above a mantle plume (a region of particularly hot mantle that has elevated the region, and which fuels the volcanic activity and geysers there). Label B marks a linear region of cooler rock that trends north to south from Washington State to northern California. This area, which lies directly east of the Cascadia Trench, is where the cool rock of the Juan de Fuca Plate—the tail end of the Farallon Plate—is plunging into the trench. Label C marks a cool area where a piece of the upper mantle has apparently detached from the base of the crust underneath the southern Sierra Nevada and Central Valley. This cool blob is presently sinking into the mantle like a great drip, and the loss of its weight from the crust above may have let the southern part of the Sierra Nevada bob up to its current stupendous heights.

a foot per mile on its east end, increasing to ten feet per mile on its west end—much less than the slope of the tableland today. Conclusion: regional uplift of the Rockies during the last five million years has tilted the Cheyenne Tableland. The difference in slope translates into nearly a half-mile of uplift of the Rockies—roughly twice what we would expect from passive, erosion-induced uplift alone and more than enough to explain the Exhumation.

McMillan's work has convinced many geologists (including me) that active uplift—probably related to that hot, buoyant mantle under the Rockies shown in figure 13.4—has played a key role in the Exhumation. But climate change may well have contributed too. Evidence indicates that the American West has become cooler and more arid over the last five million years. While it might seem strange that increased aridity would cause more erosion, arid climates, more often than not, have very focused and intense rainstorms. To put it simply, three inches of rain dropped in one big storm can do a lot more erosion than thirty inches spread out over a year. Moreover, we know that mountain glaciers first formed in the Rockies about three million years ago, showing that the mountain peaks had begun to hold more snow. We don't know whether these glaciers formed because the peaks were lifted to higher elevations or because climate change increased winter snow packs. Either way, surging spring snowmelt and more intense floods would have given Rocky Mountain rivers more power to carry out the Exhumation.

To sum up: during the last five million years, three developments have apparently all worked together to unbury the Rocky Mountains: passive uplift caused by erosion, active uplift caused by a hot, buoyant mantle beneath the Rockies, and climate change that delivered more intense floods to Rocky Mountain rivers.

. . .

So far, we've been seeking the causes of the Exhumation within the confines of the Rocky Mountains and the western Great Plains. But I suspect that the Exhumation is intimately linked to a bigger story—the death of the Farallon Plate.

From the time that we arrived at Golden Gate thirteen chapters back, the Farallon Plate—that great slab of old ocean floor, now nearly extinct, that began subducting under western North America some 140 million years ago—has figured large in the geologic events that we've explored. Now, like a movie star in celluloid death, the Farallon Plate is going to make one final dramatic gesture as it exits the scene, leaving its end mark—the Exhumation of the Rocky Mountains—on our story of the American West.

The Farallon Plate is largely gone today, a casualty of North America's westward migration. During the last twenty-eight million years, our westering continent has progressively overtopped the Farallon-Pacific Ridge—the mid-ocean ridge from which the Farallon Plate originally grew and slid east. Wherever the continent overrode the ridge, it made contact with the Pacific Plate on the other side. Whereas the Farallon and North American plates had been heading in opposite directions (so that the former dove beneath the latter), the Pacific and North American plates were going in nearly the same direction. The slight difference in their relative directions—west for the North American Plate, northwest for the Pacific Plate—resulted not in subduction at their touching edges but in side-by-side sliding. A new plate boundary was thus born—a side-by-side sliding boundary represented today by the San Andreas fault and its southward extension, the Gulf of California. If you flip back to figure 6.8, you can see these developments portrayed in map view.

Wherever North America overtopped the Farallon-Pacific Ridge, subduction ended, and the Farallon Plate was cut off from its source. That region of terminated subduction presently extends from Cape Mendocino in northern California south to Puerto Vallarta, Mexico, at the mouth of the Gulf of California. Two small, tail-end segments of the Farallon Plate remain in view on the ocean floor today. These are the Juan de Fuca Plate, which subducts down the Cascadia Trench from Cape Mendocino north to Vancouver Island, and the larger Cocos Plate, which subducts under Mexico south of the Gulf of California. (The frontispiece plates map and figure 6.8 show them both.) Both plates are about to be swallowed by their respective trenches as they follow the rest of the Farallon Plate into the mantle. When I look at them, I think of a snake's tail about to disappear down its burrow (although a beaver's tail, since it's flat, might make a better image).

To see how the termination of the Farallon Plate might relate to the Exhumation of the Rocky Mountains, look at figure 13.5. The top and middle panels in figure 13.5 will look familiar; they're reproduced from earlier chapters. In the top panel, set at thirty million years ago, the Farallon Plate has not yet been terminated. It spreads east from the Farallon-Pacific Ridge, and subducts under a region of volcanic mountains that will later become the Great Basin. Meanwhile, the Rockies are being buried in sedimentary and volcanic debris.

In the middle panel, set at fifteen million years ago, the edge of North America has overridden the Farallon-Pacific Ridge, cutting the Farallon Plate off from its source of seafloor spreading at the ridge. The detached plate sinks into the mantle, opening a gap behind the plate (a "slab window" to use technical terminology) through which hot mantle rock rises. The crust above, tethered now to the Pacific Plate on its west edge, is stretching to create the Basin and Range Province and the Great Basin—a process that continues today.

In the bottom panel, representing the last five million years, the gap behind the sinking Farallon Plate continues to open. The widening gap allows yet more hot mantle rock to well up, reaching under the Rocky Mountains and buoying them to trigger the Exhumation. The map of present-day mantle temperatures shown in figure 13.4 fits this interpretation quite well.

So what, then, can we see of the disappeared Farallon Plate—that mighty slab of seabed responsible for creating so much of the American West? Until recently, it appeared as nothing more than a fuzzy blob on seismic images of the mantle beneath North America. But more detailed seismic tomography, incorporating data from USArray, has changed that. Today, we can see the Farallon Plate quite clearly, thanks to the work of numerous geophysicists, including Karin Sigloch and colleagues. Figure 13.6, adopted from Sigloch's work, represents one of the best views yet of the Farallon Plate under North America. Notice on figure 13.6 that the plate appears to have split into

~30 million years ago

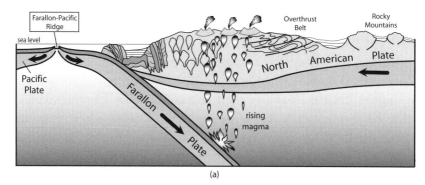

(a)

~ 15 million years ago

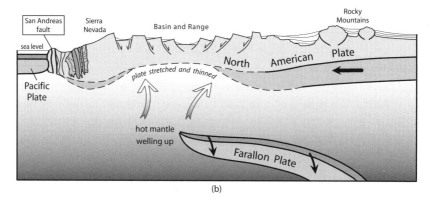

(b)

Last ~5 million years

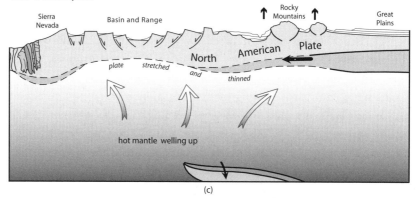

(c)

FIGURE 13.5

Time diagrams linking the Exhumation of the Rocky Mountains to the demise of the Farallon Plate. Thirty million years ago (top): Subduction of the Farallon Plate unleashes caldera volcanism across the future Basin and Range. The Rocky Mountains are buried by their own eroded debris and by volcanic ash blown eastward from the calderas. Meanwhile, the North American Plate is about to override the Farallon Plate and make contact with the Pacific Plate. Fifteen million years ago (middle): North America has overridden the Farallon Plate, cutting it off from its source at the Farallon-Pacific Ridge. →

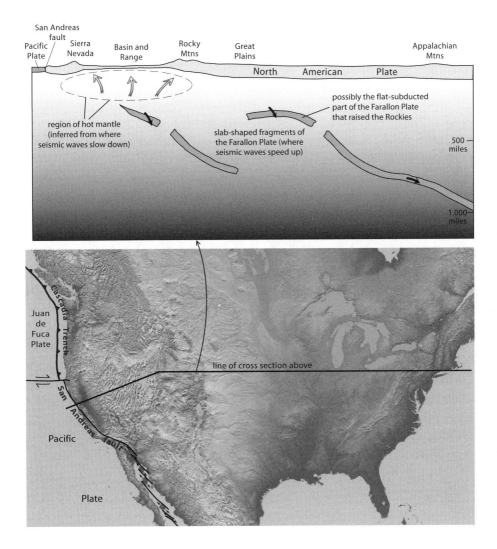

FIGURE 13.6

East–west side view (top) under North America along the line shown on the map (bottom). Seismic tomography, incorporating data from seismometers across the continent—including those in the USArray grid—has recently revealed the remains of the Farallon Plate. The diagram portrays side-view "slices" through large slabs of the plate. The plate appears to have cracked into several fragments. Notice the high fragment under the middle of the continent, which may represent the flat-subducting portion of the plate that pushed up the Rocky Mountains during the Laramide Orogeny. For clarity, I have exaggerated the topography of the land surface.

As the plate sinks, hot mantle rises through the opening gap. The crust above stretches to form the Basin and Range. The Rockies continue to be buried. Last five million years (bottom): The widening gap above the sinking Farallon Plate allows hot mantle to well up under much of the western United States, uplifting the Rocky Mountain region to trigger the Exhumation. Today, the Rocky Mountain region is also stretching east to west along the Rio Grande Rift, much like the Basin and Range in its early stages.

several fragments. (The positions of the fragments are inferred from where seismic waves speed up, indicating that they are passing through rock cooler than the surrounding mantle.) To the west of these fragments, a region of seismic wave slowing reveals the hot mantle presently welling up under the Basin and Range and the Rockies. For me, the most exciting thing about figure 13.6 is that high-floating piece of the Farallon Plate that you can see under the Great Plains. It's likely that that high fragment is the portion of the Farallon Plate that slid flat under the continent to squeeze up the Rocky Mountains during the Laramide Orogeny.

I'm back on the Gangplank, where the remnant Miocene plains still rise, as ever, to the summit of the Rocky Mountains. The tall grass swoops east, combed by the incessant wind. The Rockies arc across the western sky. There, in my book—as well as this book—is where the West begins. When, as a young man, I first crossed these plains and came to those mountains, it turned the arc of my life forever West. That was nearly three decades ago, and I didn't know, then, that I was looking at the frontier of the rest of my life. I didn't know, either, where the Rockies came from—no one really did. But science, done right, frequently circles back on itself to look at old questions through new eyes. Now, with the new eyes of seismic tomography, we can look *down* into the planet and see the great plate that did the work—the one that not only pushed up the Rockies long ago but also, in dying, exhumed them from burial and gave them back to the world (along with Coors). It lies in its mantle grave, revealing itself by the distant echo of seismic waves 300 miles beneath my feet.

NOTES

1. Dodge is credited with leading the exploring party that, in 1865, discovered the table-land route for the transcontinental railroad. He also founded Cheyenne, the capital of Wyoming.

2. A pediment, as you may recall from chapter 11, refers to the smooth surface that streams cut onto bedrock where mountain fronts retreat laterally from their adjacent valleys. Normally, pediments merge seamlessly with the surfaces of the gravel-filled valleys next door. But if erosion later scoops the gravel out of the valleys, the pediments are left hanging as step-like sub-summit surfaces high on the mountainsides.

3. Here's the math to demonstrate the point. The area of the Colorado Piedmont is approximately 8,000 square miles. Assuming a quarter-mile-deep excavation, on average, of the former Great Plains across the piedmont (based on data in McMillan, Heller, and Wing 2006), this yields an excavated volume of 2,000 cubic miles. The volume of the Grand Canyon can be estimated by treating it as triangular trough 270 miles long, 10 miles wide (on average, at the rim), and one mile deep (on average, at the apex of the trough). This yields an excavated volume of 1,350 cubic miles, or 30 percent less than the volume of the Colorado Piedmont. Both excavations—the Grand Canyon and the Colorado Piedmont—have been carried out during the last five to six million years.

4. I use five million years ago as a midrange estimate for the onset of the Exhumation. Data in McMillan, Heller, and Wing 2006 indicate that the Exhumation began six to eight million years ago in the central Rockies, and three to four million years ago in the peripheral regions.

5. Braided streams are those in which the channels split and rejoin around numerous sandbars, forming a pattern somewhat like braided hair. Nebraska's Platte River is a classic example.

DEEP TIME

Fathoming the Rock Record

> If only the Geologists would let me alone, I could do very well, but those dreadful
> hammers! I hear the clink of them at the end of every cadence
> of the Bible verses.
>
> JOHN RUSKIN, 1851

In 1655, Irish Archbishop James Ussher figured out the age of the Earth. Using a pains-taking analysis of biblical generations, Ussher determined that God had created the Earth on October 23, 4004 B.C.

The news hardly surprised anyone. Many were delighted with Ussher's precision, but it didn't change prevailing wisdom. Scripture implied an age for the Earth compat-ible with the scope of human history. Ussher simply put a number on what everyone already knew. If you wanted to understand the Earth's history in Ussher's day, you natu-rally looked to Scripture. Where else would you go? The notion that rocks can tell their own stories about the Earth, without help from the Bible, involved an intellectual revo-lution. In particular, it involved the recognition of deep time[1]—the realization that the Earth is incredibly old.

No concept has been more important for geology. As the late Steven Jay Gould once remarked, "All geologists know in their bones that nothing else from our profession has ever mattered so much." Earth processes that seem trivially slow in human time can accomplish stunning work in geologic time. Let the Colorado River erode its bed by 1/100th of an inch each year (about the thickness of one of your fingernails). Multiply that by six million years, and you've carved the Grand Canyon. Take the creeping pace at which the continents move (about two inches per year on average, or roughly as fast as your fingernails grow). Stretch that over thirty million years, and a continent will travel nearly 1,000 miles. Stretch that over a few billion years, and continents will have time to wander from the tropics to the poles and back, crunch together to assemble super-continents, break apart into new configurations—and do all of that again several times over.[2] "Time," the physicist John Wheeler once quipped, "is Nature's way of keeping everything from happening at once." Deep time, it could be said, is Nature's way of giving the Earth room for its history. The recognition of deep time may be geology's paramount contribution to human knowledge. But how do we know about deep time? How can we infer the amount of time involved in geologic events, tell the age of a rock, or know the age of the Earth? We'll begin with an example from the Great Basin and go from there into the bigger story.

Few people outside of Nevada could point to the Humboldt River on a map. But for gold rush–era pioneers headed to California, no river was more important. The Humboldt formed a lifeline of water and grass for 350 miles through the arid lands of what is now north-central Nevada. Where the Humboldt encounters the Adobe Range a few miles east of Carlin, Nevada, the river cuts a 500-foot-deep chasm called Carlin Canyon. The back-and-forth wanderings of the river between the canyon walls forced the pioneers to use the thigh-deep riverbed as their wagon road. Splashing along, they all passed the soaring outcrop shown in figure A.1, whose "mighty mass," one forty-niner wrote, "seemed to hang over our heads, and threaten to overwhelm us with its ponderous weight." In the century and a half since, this outcrop has posed for countless geology cameras and been projected in front of I-don't-know-how-many darkened classrooms—including mine. It stands as Exhibit A in our story of deep time.

Notice in figure A.1 that the rock layers on the left are standing vertically, like books on a shelf. These are sedimentary layers, mostly beds of conglomerate (solidified gravel and pebbles) and sandstone (solidified sand) called the Diamond Peak Formation. Now notice that the layers on the right tilt steeply down to the right. These are limestone layers of a younger rock unit called the Strathern Formation. Both sets of layers once laid flat, like blankets on a bed. The sequence of events that explains the tilting of these layers goes like this: Rivers flowing off now-vanished mountains in western Nevada once carried sand and gravel down to the shore of an ancient sea. The sediments stacked up as flat-lying layers to become the Diamond Peak Formation (figure A.2a). Mountain upheaval then pushed and tilted the layers up at an angle (A.2b). Erosion slowly chewed those mountains down to a flat plain, beveling off the edges of the tilted layers (A.2c). The sea returned. It flooded across the eroded plain, depositing layers of limestone that would become the Strathern Formation (A.2d). Mountain upheaval again wracked the region, further tilting both sets of layers as it pushed them up into yet more mountains (A.2e). Finally, much later, the Earth's crust began to break up to form the present-day mountains of the Basin and Range. The Humboldt River cut down as the mountains rose, carving Carlin Canyon and revealing this story (A.2f).

Push up some mountains. Cut them down. Drown the land under the sea. Push up some more mountains. Cut *them* down. Push up a *third* set of mountains, and let the Humboldt River cut through them. That, in a nutshell, is the story told by the outcrop

FIGURE A.1

View north, in winter, toward the unconformity exposed on the north wall of Carlin Canyon, Nevada. The vertically tilted layers of the Diamond Peak Formation, on the left, consist of interleaved beds of chert pebble conglomerate and sandstone of early to middle Mississippian age. The steeply tilted layers of the Strathern Formation, on the right, consist of thick beds of fossil-rich marine limestone of latest Pennsylvanian to earliest Permian age. The unconformity—the ancient erosion surface that divides the two rock formations—represents roughly 40 million years. Figure A.2 shows the sequence of events needed to explain this outcrop.

in Carlin Canyon. Central to this story is the unconformity between the two differently tilted sets of layers. "Unconformity" is the geologic term for an old, eroded land surface buried under younger rock layers.[3] Imagine the sea rising and inundating an old cobble road built in Boston before the American Revolution. Mud sifts down through the seawater and hardens into shale on top of the road. Where the shale meets the road, you've got an unconformity. In this instance, the unconformity would represent, say, 300 years—the age difference between the pre-revolution cobble road and the new-formed shale on top. Put your outspread hand across the Carlin Canyon unconformity, and your fingers span roughly forty *million* years—the time that it took to bevel down the first set of mountains (figure A.2b, A.2c) and deposit the younger layers on top (figure A.2d). What is forty million years? Enough time for a small predatory dinosaur to evolve into a bird. Enough time for a four-legged, deer-like land mammal to evolve into

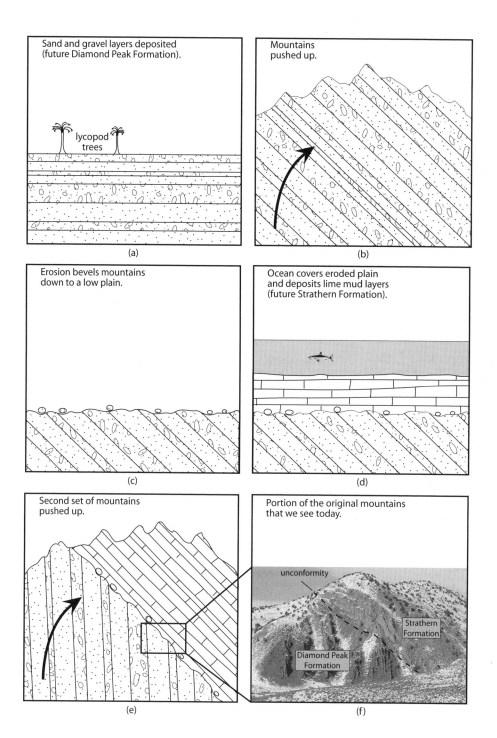

FIGURE A.2

The sequence of events needed to explain the unconformity in Carlin Canyon, Nevada.

a whale. And far more than enough time to turn an ape-like creature in eastern Africa into a big-brained biped who can marvel at such things.

While the Earth's history is written in its rocks and fossils, often it is the *absence* of rock—represented by unconformities—that testifies most clearly to the scope of deep time. Even at the Grand Canyon, arguably the world's most stunning geologic outcrop, far more time is embodied in unconformities than in rock (figure A.3). Exhibits at the Grand Canyon visitor's center compare the mile-thick stack of sedimentary layers to pages in a book of the Earth's history. But this analogy only works if you imagine a book vandalized and torn apart. Eighty percent of the Grand Canyon's stone pages have been torn out by erosion, leaving unconformities. The most profound is the so-called Great Unconformity, near the canyon's bottom (figure A.4). In some places, the Great Unconformity divides 1.7-billion-year-old rock from 550-million-year-old rock—a gap of more than one billion years. *One billion years.* I earn my salary studying the Earth and teaching its history, but I admit utter helplessness in comprehending such a span. A billion pages like those of this book would stack up more than forty miles. I had lived one billion seconds a few days before my thirty-second birthday. A tape measure one billion inches long would stretch two-thirds of the way around the Earth. Such analogies hint at what deep time means—but they don't get us there. "The human mind may not have evolved enough to be able to comprehend deep time," John McPhee once observed, "it may only be able to measure it."

.   .   .

In 1655, when Archbishop Ussher decided that the Earth had only been around since 4004 B.C., no one understood unconformities or what they implied about the scope of geologic time. It took a Scot named James Hutton, more than a century after Ussher, to figure that out.

James Hutton, a wealthy, educated gentleman farmer of late eighteenth-century Scotland, spent much of his time pondering rocks and landscapes. Wandering his native moors, Hutton observed how streams inexorably wore down the land. Such erosion, he reasoned, must be balanced by uplift of land elsewhere—otherwise the Earth would have no mountains, just low eroded plains. Hutton developed a theory (presented in 1785 in a paper that he read before the Royal Society of Edinburgh) to explain how the land surface could be renewed in the face of relentless erosion. He proposed that new land formed where ancient seabeds rose up to form mountains, pushed up by the Earth's internal heat. Rain and rivers then slowly wore down the mountains, carrying the eroded particles to the sea, where they stacked up as rock layers on the seabed. Eventually, he proposed, the restless Earth heaved these rocks out of the sea to form new sets of mountains, replacing those that erosion had elsewhere cut away. The thing that was needed for these slow cycles of renewal and degradation was *time,* and in vast amounts—spans that no one before Hutton had ever imagined. "Time, which measures everything in our idea . . . is to nature endless and as nothing," Hutton argued.

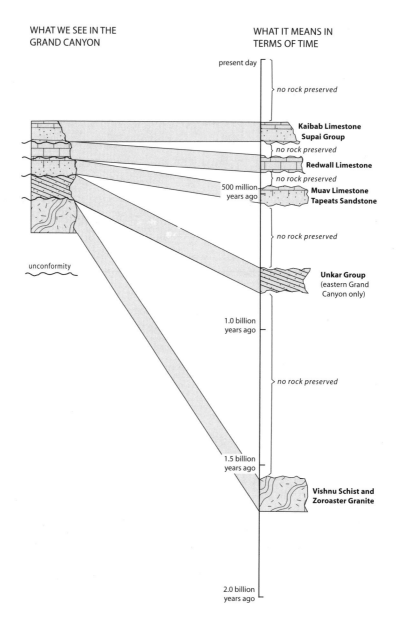

WHAT WE SEE IN THE
GRAND CANYON

WHAT IT MEANS IN
TERMS OF TIME

present day

no rock preserved

Kaibab Limestone
Supai Group

no rock preserved

Redwall Limestone

no rock preserved

500 million
years ago

Muav Limestone
Tapeats Sandstone

no rock preserved

Unkar Group
(eastern Grand
Canyon only)

1.0 billion
years ago

unconformity

no rock preserved

1.5 billion
years ago

Vishnu Schist and
Zoroaster Granite

2.0 billion
years ago

FIGURE A.3

The rocks of the Grand Canyon placed on a vertical time line. The unconformities between the rock units represent a far greater portion of geologic time than do the rocks themselves.

The reason, he suggested, is that we live on an unimaginably old Earth—one with "no vestige of a beginning,—no prospect of an end."

In 1788, three years after he presented his theory and the great age of the Earth that it implied, Hutton and his friend John Playfair (a professor of mathematics at the University of Edinburgh) were exploring Scotland's east coast by boat. They pulled ashore

FIGURE A.4

The Great Unconformity in the inner gorge of the Grand Canyon. The top photograph looks west along the gorge at the level where the Tapeats Sandstone (550 million years old) overlies the Vishnu Schist (1.7 billion years old). The unconformity between them encompasses a gap of more than 1 billion years. The bottom photograph shows a close-up of the Great Unconformity along one of the canyon's hiking trails. My friend Dan, awed by the presence of the unconformity (or perhaps just sick of my blathering about it), leans on the Vishnu for support.

Devonian
sandstone

Silurian
sandstone and shale

FIGURE A.5

The unconformity at Siccar Point on the east coast of Scotland—perhaps the most famous outcrop in the history of geology, where, in the late eighteenth century, James Hutton affirmed his theory of the Earth's great age. The layers of Silurian sandstone and shale, after being laid down flat, were tilted nearly vertical and beveled off by erosion. Layers of Devonian sandstone were then laid on top, producing an unconformity. Comparing the geometric relationships of the rock layers here with those of Carlin Canyon in figure A.1, you can see that the sequence of geologic events needed to explain both unconformities is similar.

at a wave-whipped headland called Siccar Point and stumbled upon the unconformity shown in figure A.5. Here, they realized, was an outcrop that spoke powerfully of the Earth's great age and the slow cycles of uplift and erosion that formed the legs of Hutton's theory.[4] "The mind seemed to grow giddy by looking back so far into the abyss of time," Playfair later famously wrote of Siccar Point. Archbishop Ussher's scriptural chronology gave the Earth only 5,792 years (4004 B.C. to 1788 A.D.) to do such work. "No," said Hutton, "that was not enough time." Not even close.

James Hutton didn't single-handedly transform our view of geologic time; the advancement of science is rarely a solitary effort. But his work did much to fan a growing intellectual flame. Within a few decades after Hutton, nearly all scientists who weighed the evidence came to understand that the Earth must be far older than the scope of Biblical time. Unconformities require it—as do other forms of geologic evidence. Unconformities, with Hutton's help, awakened the human mind to the scope of deep time.

But how deep, exactly? Today we estimate the Earth's age at 4.56 billion (4,560,000,000) years. Science's journey to that number arguably began with French naturalist Georges-Louis Leclerc, Comte de Buffon, in the late eighteenth century.

Buffon was, by any measure, one of the liveliest characters of eighteenth-century science. He kept wild animals in his house for study, argued that the planets had formed from a comet crashing into the Sun, and once built an experimental mirror that, by focusing the Sun's rays, could set fire to houses 200 feet away. He paid a servant a crown per day to wake him at 5:00 a.m. every morning as part of a fanatical work regime that, among other things, produced his monumental thirty-six-volume *Historie Naturelle*, in which he attempted to unite all known natural history, geology, and anthropology in a single work. In the 1770s, Buffon devised the first scientific experiments aimed at figuring out the age of the Earth. Buffon reasoned that the Earth had cooled down to its present temperature from an initially white-hot or molten state. To mimic the process, he heated up iron spheres of various sizes to near their melting point and measured how long they took to cool. He repeated the experiment with other substances, including several kinds of rock. With some scribbles of his quill to scale the experiments up to Earth-sized spheres, Buffon guessed that the Earth was about 75,000 years old.

Although absurdly young from our present perspective, Buffon's calculations—published in 1778, seven years *before* James Hutton's seminal 1785 old-Earth paper—caused a major stir. The French Catholic Church burned his books and threatened excommunication. Buffon begged forgiveness—and then proceeded to reassert his conclusions throughout his later writings. Perhaps, like a fish in a growing school, he felt protected from the attacks of biblical dogmatists by the swelling ranks of like-minded thinkers throughout Europe—a luxury that Galileo some 140 years earlier had not enjoyed.

Buffon's calculations were flawed, in part, because he assumed a linear cooling rate, when in fact hot objects cool exponentially, losing heat more rapidly at first and less rapidly later. Nearly a century later, a different error of assumption would trip up a scholar of far greater renown: the great British physicist William Thomson.

Thomson[5] was a colossus of nineteenth-century science, the type of man who electrified a room simply by walking in. The bow wave of his accomplishments pushed aside all competition, and men who were highly accomplished scientists in their own right often grew rather cowed in his presence. "I felt quite wooden beside him sometimes," the eminent physicist Hermann von Helmholtz once admitted. Thomson entered Glasgow University at the age of ten, and by the time he was sprouting respectable facial hair had produced several mathematical papers of such stunning originality that he had to publish them anonymously to protect his superiors from embarrassment. He became a professor at age twenty-two, and over the course of his long career he published some 660 papers, many of which contributed fundamentally to our modern understanding of electricity, magnetism, thermodynamics, hydrodynamics, and geomagnetism.

He also developed some seventy patents, several of which made him very rich. Suffice it to say that when Thomson weighed in on the question of the Earth's age, more than a few people took notice.

To calculate the Earth's age from its cooling rate, Thomson needed three pieces of information: the initial temperature of the Earth; the Earth's temperature "profile," meaning the rate at which it heats up going from the surface downward (what we today call the geothermal gradient); and the thermal conductivity of rock (a measure of how quickly heat travels through rock). Thomson used 7,000 degrees Fahrenheit as an estimate of the Earth's initial temperature, based on experiments melting rocks. The geothermal gradient he calculated from the rise in temperature in deep mines. The third value, thermal conductivity, came from experiments heating up and cooling down rock. In 1862, he announced his results: the Earth had formed sometime between 20 million and 400 million years ago. The wide range reflected Thomson's prudent uncertainty about the numbers he put into his calculations. Over the next three decades, however, Thomson became simultaneously more precise and more wrong. He revised his age estimates continually downward, and by 1893, he had concluded that the Earth was probably no more than 24 million years old.

The problem was that, ever since Hutton, geologists had become convinced that the Earth was at least several hundred million years old, and probably much older. The record of layered rocks and unconformities seemed to demand it. Geologists had measured how fast silt and mud layers were building up on the modern seabed and had extrapolated those rates to the great thicknesses of sedimentary rocks they saw pushed up in the Alps and elsewhere. Such thick accumulations seemed to demand much more time than Thomson's estimates. Moreover, the unconformities between the rock layers implied vastly more time even than the layers themselves. Speculations about the Earth's age so at odds with his calculations set the great Thomson aquiver with indignation. Who did these by lowly stonebreakers—as geologists were then derisively known—think they were, making their fuzzy claims in the face of his elegant mathematics? "A great reform in geological speculation now seems to be necessary!" he spluttered—and proceeded, for the rest of his life, to throw the weight of his reputation into attacks against those who spoke for an older Earth.

Thomson, we now know, was on the wrong side of history. But how did the leading scientist of his day, using precise mathematics and the best available data, come up with an estimate of the Earth's age that was wrong by *200-fold*? The answer seems to be that Thomson made a fatally wrong assumption about the inside of the Earth. He assumed that the rock of the Earth's interior is rigid, like a steel ball, and that heat therefore travels outward from the interior only through conduction—the migration of heat through stationary material. Not everyone agreed with Thomson's rigid-interior assumption. One of Thomson's assistants, an engineer named John Perry, realized that if the Earth was hot enough inside for the rock to ooze and flow slowly (visualize soft wax), then heat could move outward through convection—the migration of heat via the actual movement of a heated material. This turns out to have been a key insight, because convection moves heat much more rapidly than does conduction.[6] In 1894, Perry proved mathematically that convection could explain the Earth's present geothermal gradient even if the Earth was very old. Indeed, Perry's calculations indicated that the Earth's present geothermal gradient was compatible with an age of several *billion* years—on par with our present estimate of the Earth's age.

Perry, bursting with excitement, shared his calculations with Thomson. But Thomson seems to have ignored them, and Perry so admired his boss and mentor that he didn't push his idea too hard. What a pity. If Thomson had appreciated Perry's insight, Perry might be recognized today as a scientific hero, with Thomson sharing credit for the breakthrough. Instead, Perry is a historical nobody, and Thomson has his otherwise sterling reputation shackled to a great error. Perhaps more significantly, if Perry's ideas about convection had come out, geologists might have recognized earlier the mobility of the continents, because the notion that the Earth's interior might be slowly churning leads logically to the idea that parts of its surface might be moving.[7]

Happily, right around the time that Perry was failing to convince Thomson, the quest to figure out the Earth's age took a monumental leap forward, thanks to the discovery of radioactivity—a natural clock within rocks that ticks to a rhythm as fundamental as the atoms from which it emanates.

.   .   .

Among the ninety-two known naturally occurring chemical elements, several include varieties that are radioactive, which simply means that they change into different forms at a constant rate that we can measure. For instance, uranium-238 (one of several varieties—or isotopes—of uranium) over time changes spontaneously into a new element, lead-206 (one of several isotopes of lead). When radioactive atoms change, or "decay," little particles jet out from the atomic nuclei at high speeds. You don't want to be near high concentrations of radioactive atoms, because when these high-speed particles shoot out, they tear through your flesh like tiny bullets, shredding your tissues and blowing holes in your DNA. A young Polish émigré to France named Marie Curie didn't know this when, in 1896, her mentor Henri Becquerel asked her to investigate the strange emanations coming from some uranium he had stored in a drawer. Curie, her husband Pierre, and Becquerel shared the 1903 Nobel Prize in physics for their discovery of radioactivity. The honor would doubtless have lost some of its luster for Curie if she had known that the fatal leukemia she contracted later in life probably came from the uranium she handled. (To this day, Curie's papers and notebooks are still so toxically radioactive that they are stored in lead-lined boxes, and anyone wishing to study them must wear protective clothing.)

The key thing about radioactivity for dating rocks is that radioactive atoms change in a predictable, clock-like way. For instance, if you have a certain amount of tritium (a radioactive isotope of hydrogen), and you wait for 4,497 +/−8 days, exactly half of the tritium will have changed to helium-3; 4,497 +/−8 days is the *half-life* of tritium, and its value is unvarying within experimental measurement error (represented by the +/−8 days). As far as we can tell, nothing can change the half-life of a radioactive element—not heat, pressure, or geologic age. Half-lives appear to be as constant and reliable as any law of physics.

Tritium isn't useful for geologic dating, because its half-life is so short. But other radioactive elements have much longer half-lives. Several of these are common enough in rocks that they provide useful clocks. The basic idea is this: As a rock congeals from magma, atoms in the magma—including any radioactive ones that might be present—assemble into mineral grains. Once locked into a mineral grain, no atom can easily escape; powerful chemical bonds keep each atom firmly attached to its neighbors. As long as the mineral grain remains a closed system with no new atoms entering or leaving, the proportion of radioactive *parent* atoms in the rock compared to their *daughter* decay products provides an accurate measure of when the rock formed (figure A.6).[8] By the 1950s, geochemists had perfected methods for counting parent and daughter atoms in rock and mineral samples. Radiometric age dates have been exploding across the geologic literature ever since.

The problem for figuring out the age of the Earth using radioactive decay does not lie in the technical details of counting parent and daughter atoms. The problem is that there are no rocks left from the planet's earliest days. At some level, this should come as no surprise; it's not easy being a rock on planet Earth, what with erosional assaults by wind, rain, rivers, and glaciers. And even if a rock escapes erosion, the Earth's moving tectonic plates regularly recycle and destroy rock. Geologists have so far found no rock older than 4.28 billion years (from the Nuvvuagittuq Belt on the eastern shore of Canada's Hudson Bay). The oldest Earth materials so far discovered are a handful of tiny 4.36-billion-year-old grains of the mineral zircon, eroded from now-vanished rocks and preserved within beds of younger sandstone in Australia's Jack Hills.

The best answer we'll likely ever get for the Earth's age comes from radiometric dating of meteorites—rocky chunks left over from the solar system's formation (figure A.7). We think that all of the objects in the solar system—the Sun, the planets and their moons, and all other material, including meteorites—formed at roughly the same time from a vast, rotating cloud of cosmic debris swept together by gravity. Touch a meteorite, and you reach back to the birth of our solar system—back to a time when the Earth was an incandescent ball orbiting a newly minted Sun. Radiometric dates on meteorites consistently yield ages as old as 4.56 billion years; 4.56 billion years is, therefore, our current best estimate for the age of the Earth.

·   ·   ·

4004 B.C. versus 4.56 billion years. We've come a long way since Archbishop Ussher. But what is 4.56 billion years? Deep time is like the federal deficit; we know it's big, but we have a hard time comprehending numbers so out of scale with ordinary experience. Let the height of a six-foot-tall man represent 4.56 billion years; the thickness of a single hair

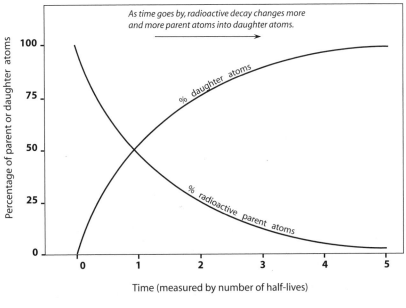

As time goes by, radioactive decay changes more and more parent atoms into daughter atoms.

% daughter atoms

% radioactive parent atoms

Percentage of parent or daughter atoms

Time (measured by number of half-lives)

Parent → Daughter	Half-life (years)	Minerals in which parent and daughter atoms are found
Uranium-238 → Lead-206	4.5 billion	zircon, uraninite
Uranium-235 → Lead-207	713 million	zircon, uraninite
Samarium-147 → Neodymium-143	106 billion	garnets, micas
Rubidium-87 → Strontium-87	48.8 billion	micas, feldspars, hornblende
Potassium-40 → Argon-40	1.3 billion	micas, feldspars, hornblende

FIGURE A.6

The radioactive atoms most commonly used in age-dating rocks. By convention, radioactive atoms are called parents, and the atoms they change (decay) into are called daughters. When a rock forms from cooling and solidification of magma, parent atoms become locked up in the rock's component minerals. No daughter atoms are present at the start. As time goes by, parent atoms decay into daughter atoms, so that the percentage of parent atoms goes down and the percentage of daughter atoms goes up. After one half-life of time, 50 percent of the original number of parent atoms remains. After another half-life, half of that 50 percent, or 25 percent of the original number of parent atoms, remains, and so on. At any point in the sequence, the ratio of parent atoms to daughter atoms, multiplied by the half-life (determined experimentally), gives the age of the rock. Some readers may wonder why carbon-14 (the basis of radiocarbon dating) is not on the list here. carbon-14 is used for dating not rock but organic material, such as wood, bone, shells, and teeth. The comparatively short half-life of carbon-14 means that radiocarbon dating only works on materials younger than about 70,000 years.

FIGURE A.7

An iron meteorite, a piece of the primordial solar system. The fingertip-sized cavities, called ablation pits, formed when chunks of the surface peeled away as the meteorite plunged through the atmosphere. The oldest meteorites yield dates of about 4.56 billion years—the value we currently take for the age of the Earth.

on his head would represent more than 200,000 years, which is longer than *Homo sapiens* has existed on Earth.[9]

"Humanity has had to endure from the hand of science two great outrages upon its naive self-love," Sigmund Freud once remarked. The first, he proposed, was the Copernican revolution, which removed us from the center of the cosmos. The second was Darwin's big idea, which linked us to all other life through a chain of common descent. A third great blow to our collective arrogance, Freud suggested—a tad immodestly—came from his own discovery of the unconscious and the inherent irrationality of the human mind. But, as Stephen Jay Gould once pointed out, Freud missed a fourth great blow— the recognition of humankind's position at the sliver end of geologic time. Perhaps, Gould suggests, "this fourth Freudian bullet" may be the most difficult to bite of all.

NOTES

1. The phrase is John McPhee's (from *Basin and Range*, 1980), but it has now become so widely used that its origin is disappearing from the collective conscious. McPhee takes this

as a high compliment. "Deep Time," he told me, "has a lot of company in the etymological cemetery, and I am really thrilled that the phrase has a stone there."

2. Here is the math to back up these statements. Grand Canyon erosion: The canyon is about one vertical mile deep, which equals 63,360 inches. Therefore, at an erosion rate of 1/100th of an inch per year, the river would cut down one mile in 6.3 million years, a value close to what geologic evidence indicates for the age of the Grand Canyon (see Karlstrom et al. 2008). Continental movement: If a continent moves 2 inches per year for 30 million years that equals 60 million inches, which is 5 million feet, or 947 miles. If a continent moves 2 inches per year for 4 billion years (somewhat less than the age of the Earth) that equals 8 billion inches, which is 667 million feet, or 126,000 miles—more than five times the circumference of the Earth.

3. To use exact terminology, the unconformity in Carlin Canyon is an angular unconformity, so called because the older rock layers (Diamond Peak Formation) were tilted at an angle and then beveled off flat by erosion before the younger layers (Strathern Formation) were laid flat on top.

4. Some accounts perpetuate the myth that Hutton's 1788 arrival at Siccar Point was his "Eureka!" moment, when he first realized that the Earth was incredibly old. In fact, he had already developed and presented his old-Earth theory three years before. The Siccar Point unconformity—along with another unconformity near Jedburgh—merely affirmed to him that he was right about the Earth's great age and gave him compelling evidence with which to convince others.

5. Upon his admission to the peerage in 1892 at the age of 68, Thomson became Lord Kelvin—the name better known to history. His contemporaries, however, knew him throughout most of his career as William Thomson, so that is what I call him here.

6. Put your hand on a warm radiator; the heat moves into your hand by conduction. Feel the air rising above a warm radiator; that heat is moving via convection of the warmed air.

7. For more on this extraordinary chapter in intellectual history, see England, Molnar, and Richter 2007. That paper also dispels a common myth about Thomson's error: that Thomson was wrong about the age of the Earth because he didn't account for radioactive heat.

8. A rock that is reheated significantly or broken down by weathering is not a good candidate for radiometric dating, not because these processes change the rate of decay, but because atoms in reheated or weathered rocks can break free and escape. If this happens, it alters the ratio of parent and daughter atoms, giving incorrect dates. Careful mapping and collection can generally eliminate such problems. Note that age dating using radioactive decay applies mostly to *igneous* rocks—rocks formed from solidification of magma. To age date *sedimentary* rocks, we ultimately depend on finding igneous rocks (particularly lava flows or volcanic ash beds) interleaved with the sedimentary layers, and then use fossils within those layers to correlate them with layers in other regions.

9. Six feet tall equals 72 inches, so each inch of height, scaled to 4.56 billion years, equals 63.3 million years. A typical human head hair is about 0.0035 inches thick. Thus, the thickness of a hair, scaled to Earth history, is 0.0035 inches × 63,300,000 years per inch = about 220,000 years.

SEEING FOR YOURSELF

This appendix will guide you to many of the key geologic sites in *Rough-Hewn Land*. The latitude–longitude coordinates, typed into a geographic database such as Google Earth, will give you the location of each outcrop or viewpoint to within a few feet. In some instances, the coordinates give the location of a parking area followed by directions for walking to an outcrop or viewpoint. Where available, I include one or more references to published field guides (citations are in the bibliography) that you can bring along for a fuller explanation.

CHAPTER 1, GOLDEN GATE

PILLOW BASALTS AT POINT BONITA, MARIN HEADLANDS

A half-mile foot trail to the Point Bonita Lighthouse reveals superb exposures of Jurassic pillow basalt scraped off the Farallon Plate as it subducted under North America's western edge during Cretaceous time. Elder 2001, Stop 2; Wahrhaftig and Murchey 1987.

Where   37.8219°, −122.5294° to park for the lighthouse trail. You'll also find excellent pillow basalt outcrops at the south end of Rodeo Beach one-quarter mile to the north. Park at 37.8252°, −122.5306°, walk west to the beach, and turn north along the beach.

PILLOW BASALTS AND CHERT BEDS AT BATTERY 129, MARIN HEADLANDS

A section of Jurassic ocean floor scraped off the Farallon Plate, from the pillow basalts of the oceanic crust up into chert beds that accumulated as deep-sea sediments on top of the pillows. Wahrhaftig and Murchey 1987.

Where   37.8282°, −122.4979°.

SERPENTINITE, SOUTH SIDE OF THE GOLDEN GATE

Spectacular blue-green rocks that were once part of the upper mantle below the oceanic crust before they were scraped off the Farallon Plate. Elder 2001, Stop 5; Moores et al. 2006, Stop 32.

Where   Along the beach about one-half mile south of the Golden Gate Bridge's south ramp. Park along Lincoln Blvd. at the small pullout on the west side at 37.8018°, −122.4774°, or anywhere else nearby, and walk west down foot trails through bluffs of blue-green serpentinite. You'll also find good exposures at 37.8103°, −122.4768° in the parking area at Fort Point under the bridge's south ramp.

CHAPTER 2, MOTHER LODE

SMARTVILLE OPHIOLITE, WESTERN SIERRA NEVADA

Jurassic pillow basalts dislodged from the deep ocean floor in the process of assembling the Western Belt of terrane rocks in the Sierra Nevada foothills (as shown in figure 2.3).

Where   Good exposures begin in road cuts at 39.2178°, −121.3326° along Highway 20 immediately east of the bridge over the Yuba River. The best pillows are at 39.2153°, −121.3248° on the north side of the highway about one-half mile east of the bridge.

NORTH STAR MINING MUSEUM, GRASS VALLEY

Exhibits of lode gold mining equipment, along with a massive cylindrical core of serpentinized gabbro from the Smartville Ophiolite. Konigsmark 2002, 170–171.

Where   39.2088°, −121.0697° in Grass Valley.

EMPIRE MINE STATE HISTORIC PARK, GRASS VALLEY

One of California's best museums for exploring nineteenth-century lode mining. Konigsmark 2002, 172–173.

Where   39.2071°, −121.0460° near Grass Valley.

FEATHER RIVER BELT, WESTERN SIERRA NEVADA

Glistening outcrops of green and black serpentinite representing the floor of an ancient ocean basin that closed as a set of volcanic islands crunched into North America's then-western edge. The belt occurs between the Eastern and Central terrane belts shown in figure 2.3. Konigsmark 2002, 182–183.

Where   Along Washington Road on the south side of the South Fork Yuba River. Exit Highway 20 at Washington Road and turn north. Good exposures begin in road cuts at 39.3349°,

−120.8034° and beyond as the road descends to the river. There are also good exposures near Highway 49 in the North Fork Yuba River canyon below the bridge over the river to Goodyear's Bar: 39.5409°, −120.8862°.

MOTHER LODE GOLD DISTRICT, WESTERN SIERRA NEVADA

The Mother Lode district runs 120 miles from Mariposa to Georgetown along the ancient Melones fault zone. I recommend Koningsmark 2002 for guiding yourself to the many sites here.

CHAPTER 3, RIVERS OF GOLD

OMEGA DIGGINS HYDRAULIC MINE

The scar of the abandoned hydraulic mine is visible from a parking area.

Where   39.3153°, −120.7532°. On the north side of Highway 20 between Grass Valley and Emigrant Gap. A paved foot trail leads from the parking area to several overlooks of the mine site and the deep canyon of the South Yuba River beyond.

Note   The parking area is not well marked or easily seen from the highway.

MALAKOFF DIGGINS STATE PARK

One of California's largest hydraulic mines; a spectacular man-made pit that is now a state park. Konigsmark 2002, 174–177.

Where   At the park visitor's center in North Bloomfield at 39.3680°, −120.8999° you can get a map to guide yourself around the hydraulic pit and see the auriferous gravels: the gold-bearing riverbed gravels washed west into California from the high Nevadaplano during Eocene time.

CHAPTER 4, A TRAVERSE ACROSS THE RANGE OF LIGHT

TERRANE ROCKS NEAR AUBURN, WESTERN SIERRA NEVADA

Distorted remains of ancient volcanic islands and slivers of deep ocean floor that subduction jammed against North America's then-western edge during Jurassic time. These rocks make up part of the Western Belt of terranes shown in figure 2.3.

Where   From Interstate 80 in Auburn, exit onto Highway 49 south and descend into the canyon of the North Fork American River. Where the road crosses the river at 38.9153°, −121.0403°, you'll find excellent exposures of green, metamorphosed volcanic terrane rocks.

LAKE COMBIE OPHIOLITE, WESTERN SIERRA NEVADA

Pillow basalts dislodged from the deep ocean floor in the process of assembling the terrane rocks of the western Sierra Nevada foothills. Moores et al. 2006, Stop 28.

Where   38.9839°, –121.0061°, in a road cut along the frontage road immediately south of Interstate 80 between Clipper Gap (exit 125) and Applegate (exit 128).

AURIFEROUS GRAVELS, WESTERN SIERRA NEVADA

Spectacular exposures of red and rust-orange auriferous gravels along Interstate 80. Konigsmark 2002, 168–169; Moores et al. 2006, Stops 26 and 27.

Where   Between Gold Run (exit 144) and Dutch flat (exit 145) on Interstate 80. A rest area one-quarter mile west of exit 144 (accessible only from the westbound lanes) offers a close-up view: 39.1790°, –120.8555°. The amphitheater-like back wall of the rest area is the headwall of the old Stewart Mine.

CONTACT BETWEEN GRANITE AND TERRANE ROCK, SIERRA NEVADA

The western edge of the Sierra Nevada Batholith where Cretaceous granite intrudes the Jurassic terrane rock of the Eastern Belt. Hanson et al. 2000, Stop 1.5.

Where   Access is on foot from the Loch Leven trailhead at 39.3089°, –120.5161° near Big Bend. Take Interstate 80 exit 166, turn northeast onto Hampshire Rocks Road, and go about one-quarter mile past the town of Big Bend to the trailhead. Follow the foot trail southwest toward Loch Leven. In about one-half mile you will pass from pure granite into the broad transition zone where the granite has intruded the terrane rock.

Note   A less-worthwhile but easier-to-get-to exposure of this contact occurs near the campground at Big Bend; see Moores et al. 2006, Stop 24.

DONNER PASS, SIERRA NEVADA

Glacial erosion, emigrant history, and the remains of the first transcontinental railroad. Konigsmark 2002, 162–167; Moores et al. 2006, Stop 23.

Where   39.3162°, –120.3263° is the location of the summit parking area along old Highway 40 at Donner Pass. Hike to any nearby high point. The view west takes in Donner Ski Bowl, a glacial-cut cirque. The view east looks down steep, glacially plucked ledges to Donner Lake. If you drive a short distance east, down the switchbacks, you'll find several pullouts with interpretive information about westbound emigrants and the construction of the transcontinental railroad. For a look at the famous "China Wall" built to support the original railroad bed, pull over at 39.3169°, –120.3196°.

CHAPTER 5, WHERE IS THE EDGE OF THE NORTH AMERICAN PLATE?

SAN ANDREAS FAULT AT WALLACE CREEK, CARRIZO PLAIN

Sideways movements along the fault have produced a right-angle dogleg in Wallace Creek (see figure 5.4). Sieh and Wallace 1987; Wallace Creek Interpretive Trail Guide, http://www.scec.org/wallacecreek/pdf/trailguide.pdf.

Where   35.2711°, –119.8271° for the location of the active channel. Walk northwest along the fault to 35.2736°, –119.8298° to see an older, now-abandoned channel.

SAN ANDREAS FAULT AT PINNACLES NATIONAL MONUMENT, CENTRAL CALIFORNIA

Sideways movements along the fault have split the Neenach volcanic field so that the two parts now lie 195 miles apart.

Where   36.4917°, –121.2095° in Pinnacles National Monument. The volcanic rocks here, on the west side of the fault, match those on the east side of the fault near Tejon Pass 195 miles to the south.

SAN ANDREAS FAULT AT LOS TRANCOS PRESERVE, SANTA CRUZ MOUNTAINS

Sideways movements along the fault have split a unit of distinctive boulders so that the two parts now lie twenty-three miles apart. Stoffer 2006, 71–72; San Andreas Fault Trail Guide for Los Trancos Open Space Preserve, http://www.openspace.org/activities/downloads/San_Andreas_Fault_Trail.pdf.

Where   37.3262°, –122.1796° for the parking area at the preserve. Leaflets at the parking area will lead you to numbered stops along the fault. The boulder beds near the parking area, on the west side of the fault, match boulder beds on the east side of the fault twenty-three miles southeast near Loma Prieta Mountain. You'll also see evidence here of fault movements during the 1906 San Francisco earthquake.

CHAPTER 6, WHERE RIVERS DIE

GENOA FAULT AND CARSON RANGE, WESTERN NEVADA

One of the best-exposed examples of an active fault in the western United States, with the exposed fault plane scored with slickenlines—parallel grooves scraped as the Carson Range lurched upward. Orndorff, Wieder, and Filkorn 2001, ch. 4; Ramelli et al. 1999.

Where   38.9854°, –119.8354°, at a small quarry along Highway 206 less than two miles south of the town of Genoa, Nevada. The exposed fault plane forms the back wall of the quarry.

Note   To preserve this outcrop, please don't hammer or climb on it.

CARSON PASS, SIERRA NEVADA, CALIFORNIA

Where the explorer John C. Frémont may have become convinced of the existence of the Great Basin as he crossed during the winter of 1844.

Where   38.6936°, –119.9872° along Highway 88.

SCARP OF THE LONE PINE FAULT, OWENS VALLEY, CALIFORNIA

A spectacular scarp whose displacement reflects movements from the March 1872 earthquake added to displacements from at least two prior earthquakes. Konigsmark 2002, 79–85; Lubetkin and Clark, 1987; Sharp and Glazner 1997, ch. 19.

Where   36.6058°, –118.0754°. Take the Whitney Portal Road west from Lone Pine for a half-mile, cross the California aqueduct, and pull out at the first dirt road on the right. Drive or walk north for 0.3 mile, bearing right after you pass through the trees.

Note   To preserve the scarp, please avoid walking on its face.

SCARP OF THE FAIRVIEW PEAK FAULT, NEVADA

A prominent scarp from the December 1954 earthquake marks the east base of Fairview Peak. Orndorff, Wieder, and Filkorn 2001, ch. 12; Slemmons and Bell 1987.

Where   39.2117°, –118.1468°. Turn south off Nevada Highway 50 at the marked sign for Fairview Peak about 40 miles east of Fallon. Drive south 5.1 miles and turn right up a small dirt track for 0.7 more miles to the parking area next to the scarp.

Note   The road is rough in places and quite steep toward the end, gaining more than 2000 feet from the highway to the parking area.

SCARP OF THE DIXIE VALLEY FAULT, DIXIE VALLEY, NEVADA

A prominent scarp from the December 1954 earthquake marks the east base of the Stillwater Range. Abbott et al. 2001; Caskey, Bell, and Slemmons 2000.

Where   39.5812°, –118.1900°. Turn north off Nevada Highway 50 onto the Dixie Valley Road about 38 miles east of Fallon. Follow the paved Dixie Valley Road north for 15 to 20 miles. You'll see the scarp along the base of the Stillwater Range to the west. Several dirt tracks lead west for a closer view.

WASATCH FAULT BY SALT LAKE CITY, UTAH

Among the many exposures of the Wasatch fault, the following three sites, together, will give you a good picture.

WARM SPRINGS SCARP A spectacular scarp displaying distinct slickenlines on the exposed fault plane.

Where   40.7982°, –111.9047°. Take Interstate 15 north from downtown Salt Lake City to exit 309 (Road 600 N), and turn east toward the Wasatch Range. Where this road intersects 300 W (also labeled Stockton Drive or Highway 89), set your odometer and turn north (left). Go past the junction for Highway 184 and get in the right lane. At 1.0 mile take a sharp right onto a gravel road alongside a chain-link fence. Follow this road into a large, gravel truck-parking area. The scarp forms the cliff at the back of the parking area.

GILBERT GEOLOGICAL PARK Two scarps of the Wasatch fault offset a glacial moraine deposited at the mouth of Little Cottonwood Canyon. Interpretive signs explain the fault, glaciations in the Wasatch Range, and the history of Lake Bonneville. Godsey et al. 2005, Stop 3.2.

Where   40.5741°, −111.7991°. From downtown Salt Lake City, take Interstate 80 east to Interstate 215 south. Exit at exit 6 and turn east toward the mountain, following the signs toward the ski areas. The road swings south along the mountain front. Continue past the intersection of Big Cottonwood Canyon Road for 2.2 miles and bear right where Wasatch Blvd forks at a traffic light. Continue for 1.1 miles more to the intersection of Wasatch Blvd and Little Cottonwood Canyon Road (9800 S). Turn right and then immediately right again into a pullout next to tiny Gilbert Geological Park.

DIMPLE DELL PARK Very close to Gilbert Geological Park, this site provides an excellent view south along the Wasatch Front to demonstrate the features of a fast-rising mountain range, including the straight range front, large triangular facets at the ends of ridges, short, steep canyons, and small alluvial fans, as shown in figure 7.5. Godsey et al. 2005, Stop 3.3.

Where   From Gilbert Geological Park, head south on Wasatch Blvd for 0.3 miles to Bell Canyon Road (E 10025 S), turn right, go 0.2 miles, and turn right on Dell Dimple Road. The road curves left and in less than 0.2 mile brings you to a parking area on the left at 40.5702°, −111.8075°. Walk 0.2 miles south along the footpath to Inspiration Point.

CHAPTER 8, WEALTH AND MAGMA

VIRGINIA CITY AND THE COMSTOCK LODE, WESTERN NEVADA

Mark Twain's adopted city and the site of the now-gutted Comstock Lode. Outcrops and road cuts around town and on the approaching roads exhibit the hydrothermal alteration characteristic of metal ore mineralization, including softening, fracturing, intrusion by quartz veins, and color alteration to mottled hues of yellow, orange, and red.

Where   39.3030°, −119.6528°.

BATES MOUNTAIN TUFF, CENTRAL NEVADA

One of many deposits of volcanic tuff hurled forth during the cataclysmic caldera eruptions of the Great Basin. Severely squashed pumice fragments testify to the extreme temperatures of the eruption clouds. Orndorff, Wieder, and Filkorn 2001, ch. 15.

Where   39.4486°, −116.7518°, the campground at Hickison Petroglyphs Recreation area along Nevada Highway 50 about 25 miles east of Austin or 45 miles west of Eureka. You'll find excellent outcrops along foot trails that lead southwest from the campground.

BINGHAM CANYON MINE, NORTHERN UTAH

The world's largest open-pit copper mine and stunning testimony to earth-moving technology.

Where 40.5355°, –112.1480°, the visitor's center at the rim of the pit. Exhibits describe the history of the mine and the geologic processes behind ore formation. No visitor access in winter.

CHAPTER 9, WATER AND SALT

MASSACRE ROCKS AND REGISTER ROCK, SNAKE RIVER VALLEY,
SOUTHERN IDAHO

An array of gigantic boulders rolled by the Bonneville Flood. Alt and Hyndman 1989, 275–276.

Where 42.6776°, –112.9867°. From Interstate 86, take exit 28 to Massacre Rocks State Park ten miles west of American Falls. Register Rock, a particularly large flood-rolled boulder inscribed with the names of westbound emigrants, lies two miles west of the park at 42.6526°, –113.0169°.

RED ROCK PASS, SOUTHERN IDAHO

The pass is the old outlet of Lake Bonneville. About 14,500 years ago, a region of loose rock gave way here to unleash the Bonneville Flood.

Where 42.3543°, –112.0494°, the location of a geologic marker along Highway 91 five miles south of Downey, Idaho.

LAKE BONNEVILLE SHORELINES

You can find Lake Bonneville shorelines scribed like bathtub rings on most of the mountains of northwestern Utah. One of the best views is from Buffalo Point on Antelope Island looking south across White Rock Bay. Godsey et al. 2005, Stop 1.1.

Where 41.0325°, –112.2562°.

CHAPTER 10, EVOLUTION'S BIG BANG

CAMBRIAN ROCKS AND FOSSILS OF THE HOUSE RANGE, WESTERN UTAH

A famous fossil-hunting locality and one of the most complete stratigraphic sections through Cambrian rocks anywhere in North America. Hintze and Davis 2003; Hintze and Robison 1987.

Where From Highway 6 in western Utah about 28 miles east of the Nevada border, turn north on Tule Valley Road at 39.0861°, –113.5622°. The House Range forms the impressive escarpment to the east. After 13.8 miles, turn right on Marjum Canyon Road and proceed to the west entrance of Marjum Canyon at 39.2455°, −113.4198°. Set your odometer here. Because of the down-to-the-east tilt of the rock layers, driving east through the canyon takes you forward in time through the Cambrian layers. After 6.2 miles, turn left on a dirt road at 39.2557°, −113.3278° and go 7.5 miles north-northeast to Wheeler Amphitheater, where the most abundant fossils occur: 39.3534°, −113.2827°. You'll see several fossil-digging operations here in the Wheeler Shale, including commercial ones.

Note This is a remote area with rough roads; go only in a sturdy vehicle with extra water, food, and spare tires.

CHAPTER 11, RANGE-ROVING RIVERS

DEVILS GATE, SOUTH-CENTRAL WYOMING

The Sweetwater River, via the process of superimposition, has cut a 300-foot-deep chasm through a granite ridge in the Sweetwater Hills.

Where 42.4466°, −107.2121°. From Wyoming Highway 220 west of Devils Gate, turn off for the Mormon Handcart Historic Center near mile marker 57. Hiking trails lead to the upstream (west) end of Devils Gate.

Note History buffs won't want to miss the chance to visit Independence Rock, off Highway 220 six miles east of Devils Gate.

WIND RIVER CANYON, CENTRAL WYOMING

The Wind River has cut a 2,000-foot-deep canyon through the Owl Creek Mountains by the process of superimposition. Blackstone 1988, 88–89; Lageson and Spearing 1991, 159–163.

Where Highway 20 between Shoshoni and Thermopolis takes you through the canyon. The south entrance to the canyon is at 43.4146°, −108.1736°.

LODORE CANYON AND SPLIT MOUNTAIN CANYON, NORTHEASTERN UTAH

John Wesley Powell, as he floated the Green River through these canyons in the Uinta Mountains during his 1869 exploring expedition, contemplated how rivers can cut through mountain ranges.

Where No road access; the approach is by boat down the Green or Yampa rivers. Lodore Canyon at the confluence of the Green and Yampa rivers: 40.5276°, −108.9853°. North end of Split Mountain Canyon: 40.4865°, −109.1805°.

SUB-SUMMIT SURFACE IN THE MEDICINE BOW RANGE, SOUTHEASTERN WYOMING

This is a long-distance vista in which you look west from the crest of the Laramie Range to the sub-summit surface on the Medicine Bow Range fifty miles away. The sub-summit surface appears as a prominent plateau on the Medicine Bow Range at about 10,000 feet elevation. Morning light is best.

Where   From Interstate 80 east of Laramie, take exit 323 and turn west, away from the rest area and the Lincoln Monument. Go 0.3 miles to a dirt road on the right (forest road 726), and follow it 1.3 miles to where the trees open up at 41.2548°, –105.4613° for a view to the west.

Note   At 8,660 feet, this site may be too snowy or muddy for access much of the year.

CHAPTER 12, UP FROM THE BASEMENT

Road cuts along nearly all range-crossing highways in the Rockies reveal good exposures of the Archean and Proterozoic rocks that form the uplifted basement cores of the Foreland Ranges. Here are just three recommended sites.

PROTEROZOIC SHERMAN GRANITE IN THE LARAMIE RANGE, SOUTHEASTERN WYOMING

Glistening crystals of 1.42-billion-year-old granite dominated by large crystals of salmon-pink feldspar weathered into a landscape of scenic boulders. Blackstone 1988, 114–115.

Where   41.1547°, –105.3817° at Venawadoo Recreation Area. From Interstate 80 between Cheyenne and Laramie, take exit 329 and turn north to Venawadoo Recreation Area. Another good stop is at the Tree Rock point of interest between the eastbound and westbound lanes near Interstate 80 mile marker 333: 41.1336°, –105.3468°.

ARCHEAN BASEMENT IN THE SOUTHEASTERN WIND RIVER RANGE, CENTRAL WYOMING

Metamorphosed sediments, schists, iron formations, and pillowed amphibolites originally deposited as underwater sediments and lava flows between 3.1 and 2.6 billion years ago. Lageson and Spearing 1991, 137–142.

Where   Along a seven-mile stretch of Highway 28 where it crests the Wind River Range near Atlantic City, beginning at 42.4873°, –108.8256° and proceeding east.

Note   If you visit this area, don't miss Red Canyon Overlook at 42.6035°, –108.6139° along Highway 28 about eleven miles east of Atlantic City and nineteen miles south of Lander. The scenic view to the northwest shows how the rising basement core of the Wind River Range tilted the overlying Paleozoic and Mesozoic strata. Blackstone 1988, 106–107.

PROTEROZOIC BASEMENT AT FLAMING GORGE IN THE UINTA MOUNTAINS, NORTHEASTERN UTAH

Sedimentary rocks of the 950-million-year-old Uinta Mountain Group form the uplifted core of the Uinta Mountains. The brick-red sedimentary layers here inspired John Wesley Powell to name this region Flaming Gorge. Blackstone 1988 (99–100).

Where   40.9147°, −109.4198° along Highway 191 at Flaming Gorge Dam twelve miles south of the Utah-Wyoming border. You'll find good views around the dam and along Highway 191 to the south where the road climbs the Uinta Mountains.

CHAPTER 13, AT THE FRONTIER

THE GANGPLANK, SOUTHEASTERN WYOMING

The Gangplank represents a remnant of the Miocene Great Plains that still reaches unbroken to the mountains. Interstate 80 and the railroad tracks use this remnant to gain the crest of the Laramie Range. Blackstone 1988, 115–117; Lageson and Spearing 1991, 41–43.

Where   41.0992°, −105.1129°. From Interstate 80 about fifteen miles west of Cheyenne take exit 345 and turn north into the truck lot immediately north of the highway. Climb the embankment on the north side to get a view. Drive west along the frontage road to see how the Gangplank narrows and then transitions onto the uplifted basement core range. An aerial photograph of the area (such as from Google Earth) will help you picture the geography while you are on the ground.

HIGH PLAINS ESCARPMENT, EASTERN COLORADO

Westbound Interstate 70 leaves the smooth, westward-rising surface of the High Plains (the remnant surface of the Miocene Great Plains) to descend into the broad, eroded bowl of the Colorado Piedmont. Chronic and Williams 2002, 81–84.

Where   Between Genoa (39.2746°, −103.5000°) and Limon (39.2648°, −103.6685°) along Interstate 70. The High Plains Escarpment—the eroded eastern rim of the Colorado Piedmont—forms the bluffs north of the highway.

SUB-SUMMIT SURFACE VIEWED FROM MOUNT EVANS IN THE FRONT RANGE, COLORADO

A mountaintop view down onto the broad Rocky Mountain sub-summit surface (also known as the Tertiary pediment), which marks the level of the pre-Exhumation Great Plains. Chronic and Williams 2002, 29–32.

Where   39.5877°, −105.6422°. From Interstate 70 about thirty miles west of Denver, take exit 240 at Idaho Springs. Go south on Highway 103 about 15 miles to the Mount Evans entrance station, and then another fifteen miles on Highway 5 to the summit parking lot at 14,150 feet. A quarter-mile hike takes you to the summit at 14,264 feet. Your view northeast, east, and southeast looks down onto the broad, rolling sub-summit surface at about 9,000 feet elevation. Once fairly smooth, the surface is now deeply incised by river valleys cut during the Exhumation.

Note   Be prepared for cold, strong winds, and thin air. The road is open only in summer and subject to weather closures. Parking can be very limited on weekends in good weather.

ACKNOWLEDGMENTS

The idea for *Rough-Hewn Land* emerged in the wake of my previous book, *Hard Road West*, but its beginnings trace back well before that. I became a western geologist while in graduate school at the University of Arizona in the 1980s. There, Karl Flessa taught me how to write and think like a scientist and demonstrated by personal example the benefits of not taking oneself too seriously. It was there, too, that I had the good fortune to take courses from William Dickinson. Although a career in paleontology was my goal at the time, Dickinson's lectures and research papers on the geologic evolution of the western U.S. planted a seed that would later sprout into my core professional interest. After spending much of the last couple of decades rambling across the continent's rough side, I realized that I wanted to tell that story. Dickinson generously helped, first with his support of my original book proposal, then with several chapter reviews, and finally with a comprehensive manuscript review. An equally far-reaching and helpful review came from the eminent geologist Eldridge Moores. In addition, Douglas MacDougall, Page Chamberlain, Pete DeCelles, Eugene Humphreys, Simon Conway-Morris, Sean Carroll, Karin Sigloch, Scott Burdick, Jim O'Connor, Frank DeCourten, Michael Collier, and Duncan Agnew all shared their expertise on individual chapters, data, interpretations, figures, and other aspects of the book. I thank, too, the many geologists, too numerous to name, whose technical publications informed the science in *Rough-Hewn Land*. I thank Jenny Wapner, my first editor at the University of California Press, for pulling from the reject pile my original proposal (a geologic road guide) and encouraging me to recast the book as a work of broader interest. My present editor, Blake Edgar, rose to the challenge of having a full book manuscript dumped unexpectedly into his lap and skillfully piloted the book through the review, revision, and publication process. The manuscript benefited from suggestions by Kate Hoffman and corrections by Rick Camp. Lynn Meinhardt did a masterful job coordinating the art program. Edward Wade oversaw the final production of the book.

Luck and circumstance have guided me into a career that pays me to do what I love best: read, write, teach, and explore the geologic world outdoors. Friends and family helped make that possible. First are the four men to whom I dedicate the book, my partners in western adventures over the years: Malcolm Meldahl, Jim Walsh, Dan Franklin, and Jim Malusa. In the early 1980s, Jim Malusa packed me—a fresh rube from the East—into his sagging Jeep and showed me around the Arizona desert. I was hooked from the moment I tripped over a cholla cactus and spilled my Mickey's Bigmouth. Dan Franklin, my friend since childhood when we battled each other with icicle swords through the New England winter, is my co-conspirator in escapades from the depths of the Grand Canyon to the heights of the Sierra Nevada. My friend Jim Walsh has one of the best eyes for the written word of anyone I know, and if anything that I have written has merit, he gets much of the credit. The rest of the credit goes to my brother Malcolm, whose heart must have sunk, sometimes, when he saw yet another chapter draft bulging in his mailbox. Yet the drafts came back each time, heavy with pencil marks, and better for them. No rule of nature says that brothers will be great friends, but we are. I thank my friends and colleagues at Mira Costa College, along with my students and their families, for their enthusiastic support. I thank my loving wife Susan for laughs and companionship, for the shared pleasure of a life in science, and for a fellow writer's empathetic ear during times when the words wouldn't assemble into working rows. My father, Edward, is a rock of support and a model for how one should live life, embracing its blessings even in the face of adversity. My mother, Eleanor, gave me her drive, curiosity, and incessant urge to learn more. Maybe those traits came packaged in genetic code, or perhaps I absorbed them as a kid, when I would watch her happily shipwrecked in a sea of books and papers, her mind glued to the question of the moment.

NOTES

I mined most of the science in *Rough-Hewn Land* from primary sources in the geologic literature, mostly from technical papers in geologic journals, field guides, and edited volumes. The human history I derived from either memoirs or reports of historical characters, or from books and papers by historians of the West. In the chapter-by-chapter notes below, I give brief descriptions of the sources that informed specific topics, followed by the sources of any quoted material. I list all sources by author and year of publication, corresponding to the bibliography. For any source not listed in the bibliography, I provide a full citation below. Numbers in parentheses indicate the page number(s) of source material.

Dedication quote:

"May your trails be . . ." Edward Abbey, 1971. *Beyond the Wall.* New York: Holt, Reinhardt, and Winston (xvi–xvii).

Quotes that follow the dedication:

"Science is nothing but . . ." Thomas H. Huxley, 1893. Collected Essays, vol. III, quoted in *Scientifically Speaking: A Dictionary of Quotations* by C. C. Gaither and A. E. Cavazos-Gaither, 2000. London, Institute of Physics Publishing (24).

"We are like a . . ." Alfred Wegener, 1929. *The Origin of Continents and Oceans.* Translated by John Biram, reprinted by Dover Publications 1966 (viii).

"Geology is not a . . ." Rick Saltus, 2009. *Why Geologists Love Beer.* Video interview at the 2009 American Geophysical Union annual meeting in San Francisco.

PREFACE

"The mountains are calling . . ." John Muir, 1873. *John of the Mountains: The Unpublished Journals of John Muir,* edited by Linnie Marsh Wolfe, 1938. Boston: Houghton Mifflin (100).

"Beyond the wall . . ." Edward Abbey, 1971. *Beyond the Wall.* New York: Holt, Reinhardt, and Winston (xvi).

"In God's wildness . . ." John Muir, 1938. *John of the Mountains: The Unpublished Journals of John Muir,* citation above (317).

"History is all explained by geography." Robert Penn Warren, quoted in *Talking with Robert Penn Warren,* edited by F. C. Watkins, J. T. Hiers, and M. L. Weaks, 1990. Athens: University of Georgia Press (26).

PART I

"For an extremely large . . ." John McPhee: McPhee 1998 (432–433)

"The formula for a . . ." Jay Trachman, radio humorist

"We learn geology the . . ." Ralph Waldo Emerson, 1883, *The Works of Ralph Waldo Emerson.* London: George Routledge and Sons (414).

CHAPTER 1

The geology of the Farallon Islands is based on Karl et al. 2001. The geology and age of the accreted terrane rocks in the Golden Gate region comes from Elder 2001 and Wahrhaftig and Murchey 1987. Information on Gustav Steinmann and the Steinmann Trinity is from Bernoulli et al. 2003 and Moores 2003. Moores 2003 also discusses Steinmann's visit with Andrew Lawson at the Golden Gate. For information on ophiolites and how they relate to the plate tectonic revolution, I relied on the work of Eldridge Moores (Moores 1970, 2003), with supplementary information from McPhee 1998 (494–511). For a general discussion of the structure of ophiolites and their worldwide distribution, see Hamblin and Christiansen 2004 (548–558). The daily volume of tidal water that goes through the Golden Gate is from Karl et al. 2001. The history of the Golden Gate Bridge is from McPhee 1998 (576–577), Niemi, Hall, and Dahne 2006, and PBS' *American Experience,* "Golden Gate Bridge," online at http://www.pbs.org/wgbh/amex/goldengate/.

"The rock of the entire . . ." and *"When struck with a hammer . . ."* Andrew Lawson, quoted in Niemi, Hall, and Dahne 2006 (160)

CHAPTER 2

My discussion of the geological assembly of California from Jurassic and Cretaceous terranes is based on Moores et al. 2006, Dickinson 2008, Moores, Dilek, and Wakabayashi 1999, Harden 2004 (179–187), and Ingersoll 2000. The geological origin of lode gold and gold production from lode mines is based on Böhlke 1999, Ash, MacDonald, and Reynolds 2001, Koningsmark 2002 (61–71, 91–227), and Hill 2006 (213–235). Information on hydrothermal precipitation at mid-ocean ridges comes from Hamblin and Christiansen 2004 (558–561). Sources of historical information are as follows: Reports of rich gold strikes are from Holliday 1999 (140, 144, and 161 footnote). The story of the widow in Downieville who sold baked pies to miners is from Holliday 1999 (153). The information about Leland Stanford is from Brands 2002 (338–342). For lode-mining methods, I relied on Grieve, Hough, and Littlejohn 1974, supplemented with material from Hill 2006 (chapter 7), Brands 2002 (234–239), Holliday 1999 (158–161), and Koningsmark 2002 (70). The information about donkeys in early lode mines is from Grieve, Hough, and Littlejohn 1974.

"As when some carcass . . ." Bancroft, H. H., 1888. *History of California, vol. VI.* San Francisco: History Company (52).

"the most astonishing mass . . ." Brands 2002 (24)

"Keep clear of there . . ." Frederick C. Stanford, January 16, 1849: Brands 2002 (85)

"The gold mania rages . . ." New York Tribune, December 11, 1848

"The whole country from . . ." San Francisco Californian, May 29, 1848

"What class ought to go . . ." Sydney Herald: Brands 2002 (56)

"I am coming back home . . ." William Swain, April 11, 1849: Holliday 1981 (63)

"plenty of gold— . . ." Edward Hargraves, 1849: Brands 2002 (61)

"disappointed and disheartened men . . ." and *"The trails and mountains were"* Alonzo Delano, 1849 (281–282)

"Notwithstanding the repeated . . ." Alonzo Delano, 1849 (372)

"Provisions, Groceries, Wines . . ." Brands 2002 (340)

"Even in the most auriferous . . ." Alonzo Delano 1849 (373)

"Quartz and granite appear . . ." Alonzo Delano 1849 (378)

"The mere thought of . . ." Brands 2002 (234)

"the best veins abound . . ." Alonzo Delano 1849 (377)

"Drip, drip, fell the water . . ." Hutchings' California Magazine, October 1857: Brands 2002 (235–237).

"Through the whole chain . . ." (figure 2.6 caption) Alonzo Delano 1849 (376)

CHAPTER 3

Alpers et al. 2005 was my main source of information about hydraulic mining in the Sierra Nevada, including data on the volume of gravel mined by hydraulic methods and the volume of mercury used in, and lost in, hydraulic mining operations. My other sources on hydraulic mining were Hough et al. 1973, Hill 2006 (243–257), Holliday 1999 (202–212, 243–253), McPhee 1998 (454–457), Wyckoff 1999, and Konigsmark 2002 (61–71, 135). The information about mercury mines in the Coast Ranges is from Harden 2004 (298–299). I used Silva 1986 for placer mining methods. For general information about gold and gold mining in California, I used Harden 2004 (186–198) and Hill 2002 (213–235). My descriptions of damage from hydraulic mining are based on Holliday 1999 (253–263, 282–299) and McPhee 1998 (469–471). Information about the legal battle and court decision against hydraulic mining comes from Holliday 1999 (282–299) and Wyckoff 1999. Evidence for the elevation of Eocene fossil riverbeds based on raindrop isotopic analysis is from Mulch, Graham and Chamberlain 2006. The timing and amount of uplift of the Sierra Nevada during the last five million years comes from Wakabayashi and Sawyer 2001, who base their interpretation on stream incision data and the westward tilt of sedimentary and volcanic layers. The story of the collapse of the ancient Nevadaplano to form the Great Basin is based on DeCelles 2004, Horton and Chamberlain 2006, Wolfe, Forest, and Molnar 1998, and Wolfe and others 1997.

"Civilization exists by geological consent." Widely attributed to William Durant, historian, 1885–1981

"a wild waste of waters . . ." San Francisco Daily Evening Bulletin, March 9, 1878: Holliday 1999 (259).

CHAPTER 4

John Muir's arrival in California is based in Muir 1914 (1–3). The story of how Josiah Whitney defined the Sierra Nevada's western edge is from Holliday 1999 (233). Konigsmark 2002 (168–169) describes the auriferous gravels along Interstate 80 near Gold Run. The age and geology of the Rattlesnake Creek Pluton along the Loch Level Trail near Big Bend is based on Hanson et al. 2000. For general information about the Sierra Nevada Batholith, I used Hanson et al. 2000 and Hill 2006 (179–209). The geology of the Donner Pass area is based on Konigsmark 2002 (162–167). For the story of the Donner party during the winter of 1846–1847, my sources were Fey, King, and Lepisto 2002 (89–98) and McLynn 2002 (326–70). For the history of emigrant wagons crossing Donner, Coldstream, and Roller passes, my sources were Fey, King, and Lepisto 2002, Graydon 1986, and Curran 1982. For the history of the Central Pacific Railroad through Donner Pass, I used Ambrose 2000 (160–162, 199–202), along with information from historic plaques posted at Donner Pass.

"'Where do you . . . ," *"It was the bloom . . . ,"* *"At my feet lay the . . . ,"* and *"After ten years of . . ."* John Muir: Muir 1914 (1–3)

"Look at the glory! . . ." John Muir, 1877: Worster, D. 2008, A Passion for Nature: The Life of John Muir. New York, Oxford University Press (205).

"At work or play . . ." John McPhee: McPhee, 1998 (487)

"Climb the mountains . . ." John Muir: Muir 1901 (56)

"found many human . . ." Wakeman Bryarly, August 21, 1849: Potter 1945 (202)

"a great many human . . ." Charles Long, 1849: Fey, King, and Lepisto 2002 (106)

"stumps from ten to . . ." A. J. McCall, September 7, 1849: McCall 1882 (79)

"Standing at the bottom . . ." Edwin Bryant, August 26, 1846: Bryant 1848 (230)

"as steep as the roof . . ." Joseph Hackney, 1849: Fey, King, and Lepisto 2002 (109)

"We came to a rim . . ." (figure 4.5 caption) Benjamin Bonney, 1846: Fey, King, and Lepisto 2002 (101)

My information about the historic 1857 Fort Tejon and 1906 San Francisco earthquakes is from Agnew 2006, Harden 2004, and Hough 2004. Rates of movement on the San Andreas fault are based on the following sources: (1) For Wallace Creek on the Carrizo Plain: Sieh and Jahns 1984, Sieh and Wallace 1987, and the Southern California Earthquake Center's Wallace Creek Interpretive Trail Guide, online at http://www.scec.org/wallacecreek/ pdf/trailguide.pdf. (2) For the Loma Prieta Mountain–Los Trancos Preserve area: Stoffer 2006 (71–72) and the Midpeninsula Regional Open Space District's San Andreas Fault Trail Guide for Los Trancos Open Space Preserve, online at http://www.openspace.org/ activities/downloads/San_Andreas_Fault_Trail.pdf. (3) For the Neenach–Pinnacles volcanic field: Harden 2004 (362–363) and Collier 1999 (86–88). (4) For the Chocolate Mountains: Harden 2004 (385–387). Data on the relative motion of the North American and Pacific plates, and for the rates of movement on the various fault systems from the San Andreas east to the Basin and Range Province, are from Antonelis et al. 1999, Atwater and Stock 1998, Bennett, Davis, and Wernicke 1999, Flesch et al. 2000, Kent et al. 2005, Oskin, Stock, and Martín-Barajas 2001, and Unruh, Humphrey, and Barron 2003. My speculations about evolving geography of the southwestern United States in the future are based, in part, on speculations by Dickinson 2009.

"similar to sea sickness . . ." D. M. Thomas, January 18, 1857; cited in Agnew 2006 (5)

PART II

"When the traveler from . . ." John Muir, 1918, *Steep Trails.* Boston: Houghton Mifflin (164)

"Supreme over all is . . ." John McPhee: McPhee 1998 (44)

"Thus far it has . . ." Cornelia Ferris, June 3, 1853: Field Guide to the Salt Lake Cutoff, Oregon-California Trails Association Annual Convention, August 18, 2005 (31)

CHAPTER 6

My discussion of the geologic history of the Great Basin and the Basin and Range Province within the overall geologic evolution of the North American Cordillera is based on the papers of William R. Dickinson (Dickinson 2002, 2004, 2006, 2008, 2009), supplemented with information from Decelles 2004 and DeCourten 2003. Evidence for the formation of the Nevadaplano and its eventual collapse to form the Great Basin is based on Mulch, Graham, and Chamberlain 2006 and DeCelles 2004. Information about the Genoa Fault is from Ramelli et al. 1999 and Orndorff, Wieder, and Filkorn 2001 (38–47). For the history of the Buenaventura River and the explorations of John C. Frémont, I relied on Frémont's own account (Frémont 1854), supplemented with information from Chaffin 2002 (194–206) and Hayes 2007 (56–87). Frémont's method for determining his elevation at Carson Pass is explained by Graham 2003. Figure 6.8 and the related discussion of how the intersection of North America with the Farallon-Pacific Ridge produced the San Andreas fault and the Basin and Range Province traces back to the work of Tanya Atwater (Atwater 1970).

"During the early 19th-century . . ." Chaffin 2002 (197)

"What gets us into trouble . . ." widely attributed to Mark Twain, but may have originated with Twain's competing humorist Josh Billings, according to William Safire in *The New York Times Guide to Essential Knowledge,* 2nd edition, 2007 (page vii).

"the best maps in my . . ." and *"the banks of the Buenaventura . . ."* John C. Frémont, December 11, 1843: Frémont 1854 (286)

"The appearance of the . . ." John C. Frémont, January 3, 1844: Frémont 1854 (300)

"With every stream I . . ." John C. Frémont, January 17, 1844: Frémont 1854 (309)

"exactly like marching up . . ." Elisha Perkins, 1849: Curran 1982 (170)

"We were sanguine to . . ." John C. Frémont, January 23, 1844: Frémont 1854 (312)

"cross the Sierra Nevada . . ." John C. Frémont, January 18, 1844: Frémont 1854 (310)

"pointed to the snow . . ." John C. Frémont, January 29, 1844: Frémont 1854 (319–320)

"We were nearly out . . ." Kit Carson, from Kit Carson Memoirs, 1809–1856: Chaffin 2002 (210)

"were obliged to travel . . ." and *"The people were unusually . . ."* John C. Frémont, February 2–3, 1844: Frémont 1854 (326)

"Our provisions are getting . . ." John C. Frémont, February 8–9, 1844: Frémont 1854 (331)

"the dividing ridge of . . ." John C. Frémont, February 14, 1844: Frémont 1854 (334)

"We encamped with the . . . ," *"2000 feet higher than . . . ,"* *"a range of mountains . . . ,"* and *"This extraordinary fact accounts . . ."* John C. Frémont, February 20, 1844: Frémont 1854 (335–336)

"deep fields of snow . . ." John C. Frémont, February 21, 1844: Frémont 1854 (336)

"rocky and snowy peaks . . ." John C. Frémont, March 27, 1844: Frémont 1854 (359)

"It had been constantly . . ." John C. Frémont, April 14, 1844: Frémont 1854 (368–369)

"The Great Basin—a term . . ." John C. Frémont, October 13, 1843: Frémont 1854 (237)

(footnote) *"Had we met such an . . ."* William Kelly, 1849: Curran 1982 (196–197)

CHAPTER 7

My data on present-day rates of movement (stretching) across the Basin and Range Province are based on Bennett, Davis, and Wernicke 1999, Thatcher et al. 1999, and Niemi et al. 2004. For information about historic earthquakes in Nevada and California, I referred to the U.S. Geological Survey Earthquake Hazards Program: Seismicity of the Western United States websites (listed in the bibliography under U.S. Geological Survey). I derived information about specific faults and historic earthquakes in the Great Basin from the following sources: (1) Lone Pine fault and the 1872 quake: Sharp and Glazner 1997 and Lubetkin and Clark 1987. (2) Pleasant Valley fault and the 1915 quake: Wallace 1987. (3) Fairview Peak and Dixie Valley faults and the paired 1954 quakes: Abbott et al. 2001, Caskey, Bell, and Slemmons 2000, and Slemmons and Bell 1987. The relationship between rates of range uplift and the geomorphic features of range fronts (i.e., the basis of figure 7.5 and the associated text) is from Burbank and Anderson 2001, chapter 10. Information about the earthquake history of the Genoa fault and the Carson Range is based on Ramelli et al. 1999 and Orndorff, Wieder, and Filkorn 2001 (38–47). Information about the earthquake history of the Wasatch fault and the Wasatch Range is from Bruhn et al. 2005, DuRoss and McDonald 2005, McCalpin and Nishenko 1996, Solomon et al. 2005, Utah Geological Survey 1996, and the Utah Geological Survey's animated Wasatch Fault Flyby, http://geology.utah.gov/utahgeo/hazards/eqfault/wfault_flyby.htm. Poulsen 1977 parses the evidence that Brigham Young's "This is the place" declaration is a myth.

"Daily it is forced . . ." Charles Darwin, 1909, *Voyage of the Beagle.* P. F. Collier and Son, (340)

"I ran out of my cabin . . ." John Muir: Muir 1914 (78)

"an arc of glowing . . ." John Muir: Muir 1914 (79)

"I kept a bucket . . ." John Muir: Muir 1914 (84)

CHAPTER 8

Statistics on Nevada metal production are from Muntean 2008a, 2008b, Price 2007, and DeCourten 2003 (162). For the history of Virginia City and the Comstock Lode, I relied on Mark Twain's personal accounts (Twain 1872), supplemented with information from James 1998 and Moreno 2000. My discussion of the geologic formation and age of Great Basin mineral deposits is based on DeCourten 2003 (162–166, 178–180), Henry 2008, Henry and Ressel 2000, Hofstra and Wallace 2006, Orndorff, Wieder, and Filkorn 2001 (203–212), Muntean 2008a, 2008b, Tingley and Pizarro 2000, and Thompson, Teal, and Meeuwig 2002. My discussion of mid-Cenozoic caldera volcanism is based on Dickinson 2006, Henry 2008, Tingley and Pizarro 2000, and DeCourten 2003 (172–178). Information about the Bates Mountain Tuff comes from Orndorff, Wieder, and Filkorn 2001 (167–176) and Wilson and Dennis 2006.

"Vice flourished luxuriantly . . ." Mark Twain: Twain 1872 (366)

"All along under the center . . ." Mark Twain: Twain 1872 (380)

"often we felt our chairs . . ." Mark Twain: Twain 1872 (310)

"Joy sat on every . . ." Mark Twain: Twain 1872 (309)

"roosting royally midway up . . ." Mark Twain: Twain 1872 (310)

(footnote) *"Over their heads towered . . ."* Mark Twain: Twain 1872 (380)

CHAPTER 9

For the geologic history of the Bonneville Flood, my sources were O'Connor and Costa 2004, O'Connor, Grant, and Costa 2002, and O'Connor 1993, supplemented with personal communications from Jim O'Connor. Information about the effects of the Bonneville Flood along the Snake River canyon is from Alt and Hyndman 1989. For the geologic story of Lake Bonneville, my main sources were Godsey et al. 2005, Oviatt 1997, and DeCourten 2003

(206–209). Information on the Pleistocene megafauna of the Lake Bonneville region is from Gilette 1996, Plataforma 2009, and University of Arizona 1998 (press release). My information about present-day Great Salt Lake comes from Chronic 1990, Gwynn 1996, and the U.S. Geological Survey Utah Water Science Center. Information about the Bonneville Salt Flats is based in part on the U.S. Bureau of Land Management Bonneville Salt Flat website, http:// www.blm.gov/ut/st/en/fo/salt_lake/recreation/bonneville_salt_flats.html. For the story of the Donner party crossing the Salt Lake Desert, I used Rarick 2008 and McLynn 2002. For firsthand accounts of the Donner crossing, I used the memoir of Eliza Donner Houghton (Houghton 1911) and the Donner party diaries, available online at http://www.donnerparty diary.com. Additional information about the nineteenth-century crossing of the Salt Lake Desert came from Bryant 1848 and Miller 1958. To guide myself along the Donner's route over the Salt Lake Desert, I used the trail guide of Roy D. Tea, http://www.scienceviews.com/ historical/hastingstrailguide.html. For the life of Lansford Hastings, I referred to Bagley 1994 and Bennett 1999 (243–247).

"Imagine a vast, waveless . . ." Mark Twain: Twain 1872 (163–164)

"2 days—2 nights— . . ." and *"This would be a heavy . . ."* Eliza Donner Houghton: Houghton 1911 (40)

"appalling field of sullen . . ." Edwin Bryant: Bryant, 1848 (175)

"the hiatus in the animal . . ." Edwin Bryant: Bryant, 1848 (173)

"only stopping to feed . . ." Reed, James, 1871, article in the *Pacific Rural Press*: Donner Party Diary for September 2, 1846 (online at http://www.donnerpartydiary.com).

"After two days . . . ," *"all who could walk . . . ,"* *"put a flattened bullet . . . ,"* and *"Anguish and dismay . . ."* Eliza Donner Houghton: Houghton 1911 (40–43)

"Here our real hardships . . ." John Breen, 1857: Rarick 2008 (75)

CHAPTER 10

The geology of Cambrian rocks in western Utah's House Range is from Hintze and Davis 2003 and Hintze and Robison 1987. My discussion of the Ediacaran fossils is based on Mc-Menamin 1998. Information about Ediacaran and Cambrian fossil sites in the Great Basin can be found in DeCourten 2003, Gaines, Kennedy, and Droser 2005, Corsetti and Haga-dorn 2003, Hagadorn and Waggoner 2000, and Hagadorn, Fedo, and Waggoner 2000. My information on the earliest biomineralized animals, or "small shelly fossils," is based on Bengston 2004. My discussion of trace fossils of the Ediacaran and Cambrian periods is based on Droser, Jensen, and Gehling 2002, Jensen, Droser, and Gehling 2004, and Martin et al. 2000. My information on the Burgess Shale and Chengjiang faunas comes from Briggs and Fortey 2005, Briggs, Erwin, and Collier 1994, Collins 2009, Conway-Morris 1998, Gould 1990, and Hagadorn 2002a, 2002b, 2002c. The discovery and naming of *Hallucigenia* and its subsequent reinterpretation comes from Conway-Morris 1998 (53–56). My fantasy dive to the Cambrian seabed and the interpretations of the animals encountered there is adopted from Conway-Morris 1998 (63–110), but see Hagadorn 2009 for an alternative interpretation of the large predator *Anomalocaris*. My discussion of what caused the

Cambrian Explosion skims the cream off a deep and complex debate. Marshall 2006 provides a thorough review. My discussion of the role of genes in the Cambrian Explosion is based on Carroll 2005 (61–79 and 137–165), Conway-Morris 1998 (147–151), and Marshall 2006 (366–369). The discussion of the hypothesis of predator-prey coevolution as a cause for the Cambrian Explosion is based on Conway-Morris 1998 (153–165), and Marshall 2006 (369–379). I borrowed the title for this chapter from Carroll 2005, chapter 6, "The Big Bang of Animal Evolution."

"To the question why . . ." and *"We should not forget . . ."* Charles Darwin: Darwin 1884 (286–287)

"from the war of nature . . ." Charles Darwin: Darwin 1884 (429)

PART III

"How I wish I . . ." James Berry Brown, July 15, 1859: *Journal of a Journey across the Plains in 1859,* edited by George A. Stewart. San Francisco: Book Club of California, 1970 (32).

"We speak of mountains . . ." John Wesley Powell: Powell 1875 (154)

"Mountains are to the . . ." John Ruskin, 1848. *Modern Painters,* v. 1, parts I and II. London: Smith, Elder and Co. (268).

CHAPTER 11

The story of John Wesley Powell's 1869 expedition down the Green and Colorado rivers is based on Powell 1895 and Stegener 1954. Powell introduced his antecedent theory for the range-roving rivers of the Rockies in Powell 1875. Archibald Marvine gives his superimposition theory for the range-roving rivers of the Rockies in Marvine 1874, the relevant portions of which are reproduced in Powell 1875. My discussion of the burial and exhumation history of the Rockies, including the superimposition history of specific rivers in the Wyoming and Colorado and the origin of sub-summit surfaces, comes primarily from Mears 1993, with supplementary information from Snoke 1993, Chronic and Williams 2002, and Lageson and Spearing 1991. The theory of mountain formation from Earth contraction originated with James Dwight Dana (Dana 1873) and was widely promoted by Eduard Suess (Suess 1885); see also Oreskes 2001. The geosynclinal theory of mountain formation traces back to James Hall and James Dwight Dana in the 1830s; see Oreskes 2001. The antecedent rivers of the Himalayas are discussed in Valdiya 1996, 1998. See Raney 2005 for a discussion of superimposition versus antecedent theory in the formation of the Grand Canyon.

"We are now in . . ." William Cornell, 1852: Moeller and Moeller 2001 (6)

"You may think you . . ." William Wilson, August 6, 1849: Holliday 1981 (204)

"This whole region of . . ." Joseph Middleton, 1849: Martin C. W., 1985, Geology and the Emigrant: Part II. *The Overland Journal,* v. 3, no. 2 (21).

"deeply the inefficiency of . . ." and *"Had I a better . . ."* Lucien Wolcott, 1850: Martin, C. W., 2009. Book review. *The Overland Journal,* v. 27, no. 3 (86).

"It is grand, it is . . ." John Edwin Banks, June 29, 1849: Scamehorn 1965 (28)

"a grand sight . . ." Lucy Cooke, 1852: Levy 1990 (11)

"It is difficult to . . ." and *"had been rent by . . ."* A. J. McCall, June 29, 1849: McCall 1882 (45)

"there was fire below . . ." Charles Parke, June 29, 1849: Parke 1989 (44)

"by volcanic force" Alonzo Delano, June 23, 1849: Delano 1854 (99)

"This is indeed wonderful . . ." Martha Missouri Moore, July 23, 1860: Munkers, R. L., 1989. Devils Gate. *The Overland Journal*, v. 7, no. 1 (4).

"We take with us . . ." John Wesley Powell, May 24, 1869: Powell 1895 (119)

"The river is running . . ." Powell, May 26, 1869: Powell 1895 (128)

"It would seem very . . ." Powell 1875 (152)

"a flaring, brilliant . . ." Powell, May 26, 1869: Powell 1895 (128)

"the boat is capsized . . ." Powell, June 8, 1869: Powell 1895 (152)

"The river . . . was running . . . ," *"The emergence of the fold . . . ,"* and *"The contraction or shriveling . . ."* Powell 1875 (152–153)

"commenced sinking their . . . ," *"gradually cut down . . . ,"* *"once extended up . . . ,"* and *"it is but recently . . ."* Powell quoting A. R. Marvine 1874: Powell 1875 (165)

"I fully concur with . . ." Powell 1875 (165)

"Every waking hour . . ." Powell 1895 (285)

"They entreat us not . . ." Powell 1895 (280–281)

CHAPTER 12

State geologic maps for Colorado (Tweeto 1979) and Wyoming (Love and Christiansen 1985) provided the information on the ages and the types of rocks that comprise the basement cores of the Foreland Ranges. For the plate-tectonic evolution of the Rocky Mountains, including both the flat subduction and post-flat subduction periods, I relied on the excellent review paper of Humphreys 2009. That paper also informed my explanation of the connection between the hydration of the lower crust during the flat subduction period and the subsequent mid-Cenozoic volcanism in the Great Basin. For an overview of how the Foreland Ranges fit into the overall geologic evolution of the North American Cordillera, I relied on Dickinson 2009. The flat-subduction theory for the Laramide Orogeny harks back to Dickinson and Snyder 1978. Other sources on the Laramide Orogeny that informed his chapter include Dickinson et al. 1988, English, Johnston, and Wang 2003, Saleeby 2003, English and Johnston 2004, and Liu et al. 2010. Saleeby 2003 discusses rock remnants left behind by the oceanic plateau that subducted under the western United States. Liu et al. 2010 argue that the oceanic plateau subducted to raise the Foreland ranges has a conjugate on the Pacific seafloor today called the Shatsky Rise. The history of the Pampean Ranges, and the evidence for flat subduction under Chile and Argentina, comes mostly from Alvarado et al. 2009 and Ramos 2009, with supplemental information from Ramos 2008, Ramos, Cristallini, and Pérez 2002, and Fromm, Zandt, and Beck 2004. For calculations of the changing buoyancy of oceanic plates caused by the subduction of oceanic plateaus, see van Hunen, van den Berg, and Vlaar 2002. The oldest known basement rocks (4.28 billion years old, from the Canadian Shield) are reported in O'Neil et al. 2008. For general information about the Colorado Rockies, see Chronic and Williams 2002, and for general information about the Wyoming Rockies, see Lageson and Spearing 1991, and Blackstone 1988.

CHAPTER 13

The history of deposition and exhumation of the Great Plains is well summarized in Maher, Englemann, and Shuster 2003. The history of Indian conflicts with the Union Pacific Railroad on the Great Plains is based on Ambrose 2000 and Bain 1999. For the causes of the Exhumation (uplift theory versus climate change theory), I used McMillan, Heller, and Wing 2006 for a perspective that favors the uplift theory, and Pelletier 2009 for a perspective that favors the climate change theory. The argument for uplift based on the slope of the Cheyenne Tableland comes from McMillan, Angevine, and Heller 2002. For an argument for climate change based on fossil plants, see Wolfe, Forest, and Molnar 1998 and Gregory and Chase 1992. For a rebuttal to that argument, see McMillan, Heller, and Wing 2006. The geology of the Gangplank is summarized by Lageson and Spearing 1991 and Blackstone 1988. The geologic story of the Colorado Piedmont is based on Chronic and Williams 2002. For information about the Rocky Mountain sub-summit surface (the pediment on the eastern face of the Front Range that marks the level of the former Great Plains), see Anderson et al. 2006 and Chronic and Williams 2002. (Chronic and Williams refer to this sub-summit surface as the "Tertiary pediment.") For the geology of the Front Range, I referred to the geologic map of Tweeto 1979. For general information about the geology of the Great Plains, I referred to Trimble 1980. I derived my information about the USArray seismometer grid from the US-Array website (http://www.usarray.org) and from Courtland 2009. The map of mantle temperatures under the western United States (the basis of figure 13.4) is based on Burdick et al. 2008, supplemented with personal communications from Scott Burdick. My information about the Farallon Plate under North America today (the basis of figure 13.6 and the associated discussion) comes from Sigloch, McQuarrie, and Nolet 2008, supplemented with personal communications from Karin Sigloch. For a discussion of the behavior of the mantle under the western United States, see Zandt and Humphreys 2008.

"The Indian's first demand . . ." General John Pope, 1866: Ambrose 2000 (173)

"We've got to clean . . ." General Grenville Dodge, 1867: Ambrose 2000 (223)

"Until they are exterminated . . ." General Grenville Dodge, 1865: Bain 1999 (232)

"The more we can kill . . ." General William T. Sherman, 1867: Ambrose 2000 (223)

"We must act with . . ." General William T. Sherman, December 28, 1866: Bain 1999 (311)

APPENDIX I

Archbishop James Ussher published his estimate of 4004 B.C. for the Earth's creation in his book *Annals of the World* in 1658 (available at the Internet Archive: http://www.archive.org/details/AnnalsOfTheWorld). Information on the age and stratigraphy of the rocks in Carlin

Canyon comes from Smith and Ketner 1975, Stewart and Carlson 1978 (text to geologic map of Nevada, 53–57), and Trexler et al. 2004. The relative proportions of unconformities versus rock in representing geologic time in the Grand Canyon comes from Marshak 2008, figure 12.10 (374). For James Hutton's story, I used Hutton's 1788 paper, along with Carruthers 1999, Gould 1987 (61–99), and Macdougall 2008 (7–12). Buffon's story comes from Richet 2007 (132–136) and Bryson 2003 (75–76). William Thomson's story comes from England, Molnar, and Richter 2007, Macdougall 2008 (12–17), Bryson 2003 (76–78), Richet 2007 (202–208, 219, 281), and Albritton 1986 (175–185). The story of John Perry's insight about convection is based on England, Molnar, and Richter 2007. Marie Curie's story comes from Macdougall 2008 (21–35) and Bryson 2003 (109–111). O'Neil et al. 2008 report the oldest Earth rocks so far discovered (4.28 billion years, from the Canadian Shield), and Wilde et al. 2001 report the oldest zircon grains so far discovered (4.36 billion years, from western Australia). My discussion of geologic dating of meteorites is derived from Macdougall 2008 (101–122). My concluding thoughts regarding Sigmund Freud are adopted from Gould 1987 (1–3) and Gould 1996 (17–29).

"If only the Geologists . . ." John Ruskin, 1851, quoted in John D. Rosenberg, *The Darkening Glass: A Portrait of John Ruskin's Genius*. New York: Columbia University Press, 1980 (30).

"All geologists know in . . ." Stephen Jay Gould: Gould 1987 (64)

"Time is Nature's way . . ." widely attributed to the theoretical physicist John Archibald Wheeler

"mighty mass seemed to . . ." G. C. Cone, September 3, 1849: Brock 2000 (93)

"The human mind may . . ." John McPhee: McPhee 1998 (90)

"Time, which measures . . ." James Hutton: Hutton 1788 (215)

"no vestige of a . . ." James Hutton: Hutton 1788 (304)

"The mind seemed to . . ." John Playfair, 1803: Carruthers 1999 (86)

"I felt quite wooden . . ." Hermann von Helmholtz: Bryson 2003 (76)

"A great reform in . . ." William Thompson: Richet 2007 (208)

"Humanity has had to . . ." Sigmund Freud: Gould 1987 (1)

"fourth Freudian bullet" Stephen Jay Gould: Gould 1996 (19)

GLOSSARY

ACCRETION (OF TERRANES) The process, usually associated with subduction, of adding terranes (regions of imported rock) onto one another or onto the edge of a continent. Terrane accretion causes continents to grow at their edges and is responsible for assembling much of western North America, including California (*see* terrane).

ADIT A horizontal passage from the surface into a mine.

ANDESITE An igneous rock formed from solidified lava, intermediate in mineral and chemical composition between basalt and rhyolite, and dominated by small crystals of plagioclase feldspar and amphibole. The same magma forms a rock called diorite if it cools slowly underground rather than erupting.

ANTECEDENCE The process whereby a preexisting river cuts through a mountain or other high topographic feature as the feature rises (*see* superimposition).

ASH (VOLCANIC) Tiny particles of solidified magma, usually with a glassy texture, ejected from volcanoes during violent eruptions and formed as escaping gases blast the magma into a fine particulate spray that solidifies upon contact with the atmosphere (*see* tuff).

AURIFEROUS GRAVELS Gold-bearing river gravels exposed throughout much of the western Sierra Nevada, representing the channels of fossil rivers that flowed west from Nevada to the Pacific Ocean during Eocene time, before ancient highlands in Nevada collapsed to form the Great Basin (*see* Nevadaplano).

BASALT A common black volcanic rock formed from solidified lava and composed of sand-sized crystals, mostly plagioclase feldspar, pyroxene, and olivine. The same magma that forms basalt becomes a rock called gabbro if it cools slowly underground rather than erupting.

BASEMENT The oldest rocks of the continental crust, typically occurring beneath younger volcanic and/or sedimentary formations except where they have been pushed up and exposed in mountain ranges. Basement rocks typically consist of metamorphic and igneous rocks of Archean and Proterozoic age that were assembled, in part, by tectonic collisions between ancient volcanic island arcs and other crustal fragments.

BASEMENT-CORED UPLIFT A mountain formed of basement rock that has been pushed upward, usually by sideways compression along large thrust faults. The Rocky Mountain Foreland Ranges of the western United States and the Pampean Ranges of Argentina are examples.

BASIN AND RANGE PROVINCE A region of north–south-oriented, fault-bounded mountains separated by sediment-filled valleys (basins), encompassing all of Nevada and portions of Wyoming, Idaho, Oregon, Utah, Arizona, New Mexico, California, and northern Mexico. Began forming about twenty million years ago by east–west stretching of the crust and continues to form today.

BATHOLITH An igneous mass of many fused plutons with an exposed area of at least forty square miles; generally long and narrow, and often composed of granite. Batholiths form from solidification of magma deep underground and are thus exposed only where overlying rocks have eroded away. The Sierra Nevada Batholith is an example.

BILATERAN An animal whose body divides into mirror-image halves. Nearly all bilateran animal groups appeared during the Cambrian Explosion and have comprised the vast majority of animal life ever since (*see* phylum, Cambrian Explosion).

BLACK SMOKER A hot-water vent on the ocean floor that emits a black cloud of metal particles precipitating from the water (*see* hydrothermal vent).

BONNEVILLE FLOOD A colossal flood unleashed when Utah's Lake Bonneville drained through Red Rock Pass into the canyon of the Snake River about 14,500 years ago (*see* Lake Bonneville).

CALDERA A large, circular depression or crater, often miles across, formed by a violent volcanic eruption followed by collapse of the ground surface.

CAMBRIAN EXPLOSION The geologically rapid evolution of diverse forms of bilateran animals between about 525 to 510 million years ago that produced most of the animal phyla on Earth today (*see* phylum).

CHERT A hard, dense, glassy rock composed of amorphous silica, commonly (although not exclusively) forming where siliceous microorganisms accumulate on the deep ocean floor.

CIRQUE The head of a glacial valley, usually with a curving, amphitheater-like shape and having steep slopes along the upper edges and a flat or hollowed-out base, sometimes occupied by a lake.

CRUST The outermost layer of the Earth, consisting of either continental crust (twenty to forty miles thick and mostly of granitic composition) or oceanic crust (up to seven miles thick and mostly of basaltic composition).

CRYSTAL A solid form of a chemical element or combination of elements with an ordered, repetitive atomic structure. The consistent geometry of angles and plane faces seen in crystals reflects the underlying atomic structure (*see* mineral).

DIORITE An igneous rock intermediate in mineral and chemical composition between granite and gabbro, dominated by plagioclase feldspar and amphibole, and formed by magma cooling slowly underground. The same magma forms a rock called andesite if it erupts as lava onto the Earth's surface.

EDIACARAN FAUNA A group of enigmatic animals that lived about 570 to 540 million years ago, shortly before the Cambrian Explosion. Ediacaran animals represent the earliest abundant animal life on Earth, but their evolutionary affiliations with later animal groups remain enigmatic (*see* Cambrian Explosion).

EXHUMATION OF THE ROCKY MOUNTAINS The uncovering by river and wind erosion of the once deeply buried Rocky Mountains to create the rugged landscape that we see today, where rivers commonly cut through mountain ranges.

FARALLON PLATE A large oceanic plate that once filled much of the Pacific Ocean basin, but which is now mostly extinct due to its subduction under North America's western edge during the past 140 million years. Subduction of the Farallon Plate created much of the topography and structure of the North American Cordillera.

FAULT A planar or gently curved fracture in the crust where the rocks on either side have shifted measurably. The energy released when a fault shifts produces earthquakes.

FAULT SCARP A cliff formed by movement along a fault during an earthquake; represents the exposed surface of a fault that penetrates deep into the crust.

FORELAND RANGES The classic Rocky Mountains of Wyoming and Colorado, along with parts of adjacent Montana, Utah, and New Mexico. Formed during the Laramide Orogeny of eighty million to forty-five million years ago, when large blocks of continental basement rock squeezed upward along thrust faults (*see* basement-cored uplift).

GABBRO A black or black-green igneous rock composed of large crystals of mostly plagioclase feldspar, pyroxene, and olivine. Formed where magma cools slowly underground. The same magma that forms gabbro makes basalt if it erupts onto the Earth's surface as lava.

GEOTHERMAL GRADIENT The increase in temperature with depth in the Earth. Typically ranges between 75 degrees and 175 degrees Fahrenheit per vertical mile in the upper few miles of the crust.

GRANITE Broadly defined, granite is a light-colored igneous rock composed mostly of large crystals of quartz, pinkish orthoclase feldspar, white or gray plagioclase feldspar, and muscovite mica, along with a scattering of darker crystals of biotite mica and amphibole. Forms several miles underground from the solidification of silica-rich magma. Granite-like rocks form most plutons and batholiths.

GREAT BASIN A region of internal river drainage within the Basin and Range Province, occupying most of Nevada, half of Utah, and portions of California, Oregon, Idaho, and Wyoming (*see* Basin and Range Province).

HYDROTHERMAL Any process or activity involving high-temperature water within the crust, especially the alteration and emplacement of minerals and the formation of hot springs and geysers (*see* ore, vein).

HYDROTHERMAL VENT A stream of intensely hot water formed where seawater percolates into fractures in the hot rock at mid-ocean ridges and then reemerges (*see* black smoker).

ICE AGES An informal term for the glaciated periods of the Pleistocene Epoch (1.8 million to 10,000 years ago), when the Earth was periodically cooler than usual and large areas of the continents were periodically covered by ice sheets.

IGNEOUS ROCK Any rock formed by solidification of molten rock, either underground or on the Earth's surface.

ISOSTATIC ADJUSTMENT The process whereby areas of the Earth's crust float in a buoyant state in the denser rock of the mantle beneath, rising up or sinking down to achieve equilibrium.

ISOTOPE Varieties of a given chemical element distinguished by different numbers of neutrons in the atomic nucleus. Some isotopes are radioactive (*see* radioactivity, radiometric dating).

LAKE BONNEVILLE A lake that occupied much of northwestern Utah, and sometimes parts of adjacent Nevada and Idaho, between about 30,000 and 10,000 years ago, reaching its maximum size about 16,000 to 14,500 years ago, when it had an acreage roughly equal to that of Lake Michigan (*see* Bonneville Flood).

LARAMIDE OROGENY A mountain-building episode from Late Cretaceous to Eocene time (eighty million to forty-five million years ago) that squeezed blocks of deep basement rock upward in Montana, Wyoming, Utah, Colorado, and New Mexico to form the Rocky Mountain Foreland Ranges (*see* foreland ranges).

LODE A mining term referring to a vein or a concentrated zone of veins (*see* vein).

MAGMA Molten rock material formed within the Earth; becomes igneous rock upon cooling and solidification. Magma that erupts onto the surface is called lava.

MANTLE The 1,800-mile-thick region between the Earth's crust and core, forming more than three-quarters of the volume of the Earth.

METAMORPHIC ROCK A rock formed by the alteration, in a solid state, of a preexisting rock by heat and/or pressure and/or fluid interactions deep underground.

MID-OCEAN RIDGES Broad, continuous ridges on the floors of all the major ocean basins, several hundred miles to more than 1,000 miles across, with rift valleys running down the centers. The seafloor spreads from the rift valleys, forming new oceanic crust through the eruption of basalt lava (*see* seafloor spreading).

MINERAL Any naturally occurring inorganic crystalline substance possessing an orderly crystalline structure and a well-defined chemical composition (*see* crystal). Aggregations of minerals form rocks.

MOTHER LODE (OF CALIFORNIA) The major lode gold mining district in California's western Sierra Nevada foothills; forms a 120-mile-long region of concentrated gold-bearing quartz veins (lodes) that mostly follows the suture line between oceanic terranes of Jurassic age.

NEVADAPLANO A highland plateau, perhaps similar to today's Altiplano of the Bolivian Andes, that, until about twenty million years ago, existed roughly where Nevada is today and which collapsed to form the Great Basin (*see* Great Basin, auriferous gravels).

NORTH AMERICAN CORDILLERA The belt of mountains that stretches unbroken from Alaska to Central America along the west side of the North American continent.

OCEANIC PLATEAU A broad region of the ocean floor made up of particularly thick accumulations of basalt lava.

OCEANIC TRENCH A deep, linear depression on the ocean floor formed by subduction, where an oceanic plate bends down underneath an adjacent plate to plunge into the Earth's interior (*see* subduction zone).

OOZE Sediment consisting of at least 30 percent microscopic skeletal remains of tiny planktonic (floating and drifting) organisms. Oozes commonly accumulate on the deep seabed far from land.

OPHIOLITE Rock of the deep ocean floor now on land, usually emplaced during subduction or continental collision.

OPHIOLITIC SUITE A section of the oceanic crust and upper mantle that typically includes, in vertical order from top to bottom, deep sea sediments, pillow basalt, sheeted dikes, gabbro, and peridotite. The peridotite may be altered to serpentinite (*see* serpentinite).

ORE A mineral deposit, usually of metal-bearing minerals, that can be mined at a profit.

OROGENESIS The tectonic processes that collectively result in the formation of mountain belts, typically involving faulting, folding, metamorphism, and magma generation on a large scale.

PANGAEA The supercontinent that began to break up about 200 million years ago, leading to the opening of the Atlantic Ocean and the present distribution of the continents.

PERIDOTITE An igneous rock, typically green or black-green, that is composed mostly of olivine with smaller amounts of pyroxene and amphibole and little or no feldspar. Peridotite is the main rock of the upper mantle.

PHYLUM (PLURAL: PHYLA) A category of zoological classification below kingdom and above class encompassing groups of organisms that share a common body design that reflects their shared evolutionary ancestry.

PILLOW BASALT (PILLOW LAVA) Bulbous, pillow-shaped lava formations formed where basalt lava erupts underwater. Most of the ocean floor beneath a veneer of younger sediment layers is made of pillow basalt produced by seafloor spreading at mid-ocean ridges.

PLACER A deposit formed where the agitating motion of streams or waves concentrates heavy mineral grains. Placers are sources of gold, diamonds, platinum, and other valuable minerals. Placer gold in the streambeds of the western Sierra Nevada foothills triggered the California gold rush of 1849.

PLATE TECTONICS The theory, confirmed by abundant evidence, that the Earth's outer rocky shell is broken up into several dozen individual plates, fifty to one hundred miles thick, that move and interact to produce earthquakes, volcanoes, and most of the major geographic features of the planet.

PLUTON A mass of igneous rock, often bulbous in shape and several miles across, formed where magma cools and solidifies deep underground. Plutons can be exposed to view where overlying rocks have eroded away, such as in uplifted mountains, including California's Sierra Nevada. Most plutons are composed of granitic rock (*see* batholith).

PYROCLASTIC FLOW (PYROCLASTIC CLOUD) An intensely hot mixture of volcanic ash, pumice, and gasses that travels like an avalanche down the sides of a volcano and sometimes great distances beyond. Where pyroclastic flows settle and solidify, they form rock called tuff.

RADIOACTIVITY (RADIOACTIVE DECAY) The process whereby an unstable atomic nucleus undergoes fission or releases particles to transform itself into a new chemical element. By convention, radioactive atoms are called "parents," and the products of their transformation (decay) are called "daughters" (*see* radiometric dating).

RADIOMETRIC DATING The science of age-dating rocks and geologic events by measuring the ratio of radioactive parent atoms to their daughter products (*see* radioactivity).

RHYOLITE A volcanic rock, light in color and rich in silica, composed of small mineral grains or volcanic ash. The same magma forms granite when it cools and solidifies deep underground rather than erupting at the Earth's surface.

SCHIST A common metamorphic rock (meaning a rock changed from a preexisting state by heat and/or pressure) characterized by distinctive layering formed from the parallel arrangement of flat mineral grains, particularly micas.

SEAFLOOR SPREADING The mechanism whereby new ocean floor is created at mid-ocean ridges as two oceanic plates diverge and magma wells up to fill the gap.

SEDIMENTARY ROCK A rock formed from eroded pieces of preexisting rocks that have been transported by wind, water, or ice and then deposited and cemented together. Also, any rock formed from particles of biological skeletons or shells, or chemically precipitated out of water.

SEISMIC BELT A long, narrow region with a distinctly high frequency of earthquakes.

SEISMIC TOMOGRAPHY A technique for analyzing global seismic waves to create two-dimensional and three-dimensional images of the Earth's deep interior based on variations in seismic wave velocities.

SEISMIC WAVES Wave-like movements of the Earth's surface and interior produced during earthquakes.

SEISMOGRAPH A ground-motion instrument used to detect and measure seismic waves.

SERPENTINITE A rock composed mostly of serpentine minerals, which in turn form largely from hydrothermal alteration of the minerals in peridotite—the main rock of the Earth's upper mantle. Serpentine minerals are commonly greenish and soft, with a greasy or silky luster and a smooth, soapy feel (*see* ophiolite, peridotite).

SINK A bowl-like valley between mountain ranges where a river ends. May contain saline lakes during times of high river flow but are often dry and coated with mud and salt. Common in the valleys of the Great Basin.

STRATA Layers of sedimentary rock originally laid down horizontally but which may be later faulted, tilted, or folded by tectonic movements.

SUBDUCTION ZONE The region where a moving oceanic plate bends down beneath an adjacent plate to form an oceanic trench. Characterized by frequent earthquakes, and volcanism in the adjacent volcanic arc (*see* volcanic arc).

SUB-SUMMIT SURFACE A step-like surface on the mountainsides of the Wyoming and Colorado Rockies, often several miles wide and typically at elevations of 9,000 to 11,000 feet, that in some areas may reflect the maximum burial level of the Rockies prior to the Exhumation (*see* Exhumation of the Rocky Mountains).

SUPERIMPOSITION (sometimes called SUPERPOSITION) The process whereby rivers establish paths on flay-lying sedimentary layers that cover preexisting mountain ranges and then cut down through the buried mountains beneath (*see* antecedence).

SUTURE A zone of faults and fractures that marks the line along which two terranes have collided and joined (*see* terrane).

TECTONICS The study of the processes and forces that cause movement and deformation of the Earth's crust on a large scale (*see* plate tectonics).

TERRANE A block of the Earth's crust, bounded by faults, whose geologic history is distinct from adjacent crustal blocks, often because it has traveled from far away. Continents grow by the accretion of terranes to their edges (*see* suture, ophiolite).

THRUST FAULT A fault in which the rock above the fault has moved up and over the rock below, usually in response to sideways compression. A common feature in the mountains of the American West.

TUFA A form of limestone (calcium carbonate) forming rough, porous encrustations along ancient lakeshores and around freshwater springs.

TUFF A rock composed of pyroclastic particles, meaning particulate material blown into the air during a volcanic eruption (*see* ash).

UNCONFORMITY A surface of contact between two bodies of rock representing an interval when no rock formed or when the intervening rock was removed by erosion. In most areas, unconformities represent more geologic time than do the rocks themselves.

VEIN A deposit of minerals that fills in a fracture or a fault, often of hydrothermal origin and sometimes containing valuable concentrations of metals like gold, silver, or copper (*see* lode).

VOLCANIC ARC A line of active volcanoes that parallels an oceanic trench where subduction is taking place. Examples include the Andes Mountains and the Cascade Range.

BIBLIOGRAPHY

Abbott, R.E., Louie, J.N., Caskey, S.J., and Pullammanappallil, S., 2001. "Geophysical Confirmation of Low-Angle Normal Slip on the Historically Active Dixie Valley Fault, Nevada." *Journal of Geophysical Research*, 106, B3, 4169–4181.

Agnew, D.C., 2006. Reports of the Great California Earthquake of 1857. University of California San Diego: Scripps Institution of Oceanography Technical Report. Retrieved from: http://escholarship.org/uc/item/6zn4b4jv.

Albritton, C.C., Jr., 1986. *The Abyss of Time*. Los Angeles: Jeremy P. Tarcher.

Alpers, C.N., Hunerlach, M.P., May, J.T., and Hothem, R.L., 2005. "Mercury Contamination from Historical Gold Mining in California." *U.S. Geological Survey Fact Sheet 2005-3014*, 1–6.

Alt, D., and Hyndman, D.W., 1989. *Roadside Geology of Idaho*. Missoula, MT: Mountain Press.
———, 2000. *Roadside Geology of Northern and Central California*. Missoula, MT: Mountain Press.

Alvarado, P., Pardo, M., Gilbert, H., Miranda, S., Anderson, M., Saez, M., and Beck, S., 2009. "Flat-Slab Subduction and Crustal Models for the Seismically Active Sierras Pampeanas Region of Argentina." *Geological Society of America Memoirs*, v. 204, 261–278.

Ambrose, S.E., 2000. *Nothing Like It in the World: The Men Who Built the Transcontinental Railroad, 1863–1869*. New York: Simon and Schuster.

Anderson, R.S., Riihimaki, C.A., Safran, E.B., and MacGregor, K.R., 2006. "Facing Reality: Late Cenozoic Evolution of Smooth Peaks, Glacially Ornamented Valleys, and Deep River Gorges of Colorado's Front Range." *Geological Society of America Special Paper 398*, 397–418.

Antonelis, K., Johnson, D.J., Miller, M.M., and Palmer, R., 1999. "GPS Determination of Current Pacific–North American Plate Motion." *Geology*, April 1999, v. 27, no. 4, 299–302.

Ash, C.H., MacDonald, R.W., and Reynolds, P.R., 2001. "Chapter Eight: Other Significant Gold-Quartz Vein Deposits." In: Ash, C.H., MacDonald, R.W., and Reynolds, P.R., 2001, "Relationship between Ophiolites and Gold-Quartz Veins in the North American Cordillera." *Ministry of Energy and Mines, Geological Survey of British Columbia*, Bulletin 108, 81–97.

Atwater, T., 1970. "Implications of Plate Tectonics for the Cenozoic Tectonic Evolution of Western North America." *Bulletin of the Geological Society of America*, v. 81, 3513–3536.

Atwater, T., and Stock, J.M., 1998. "Pacific–North America Plate Tectonics of the Neogene Southwestern United States: An Update." *International Geology Review*, v. 40, 373–402.

Bagley, W., 1994. "Lansford Warren Hastings: Scoundrel or Visionary?" *The Overland Journal*, v. 12, no. 1, 12–26.

Bain, D.H., 1999. *Empire Express: Building the First Transcontinental Railroad*. New York: Penguin Books.

Bengston, S., 2004. "Early Skeletal Fossils." In: Lipps, J.H., and Waggoner, B.M., eds., "Neoproterozoic-Cambrian Biological Revolutions. *Palentological Society Papers 10*, 67–78.

Bennett, C.L., 1999. *Roadside History of Utah*. Missoula, MT: Mountain Press.

Bennett, R.A., Davis, J.L., and Wernicke, B.P., 1999. "Present-Day Pattern of Cordilleran Deformation in the Western United States." *Geology*, April 1999, v. 27, no. 4, 371–374.

Bernoulli, D., Manatschal, G., Desmurs, L., and Müntener, O., 2003. "Where Did Gustav Steinmann See the Trinity? Back to the Roots of an Alpine Ophiolite Concept." In: Dilek, Y., and Newcomb, S., eds., "The Ophiolite Concept and the Evolution of Geologic Thought." *Geological Society of America Special Paper 373*, 93–111.

Blackstone, D. L., Jr., 1988. *Traveler's Guide to the Geology of Wyoming* (2nd ed.). Geological Survey of Wyoming Bulletin 67. Laramie: University of Wyoming.

Böhlke, J.K., 1999. "Mother Lode Gold." In: Moores, E.M., Sloan, D., and Stout, D.L., eds., "Classic Cordilleran Concepts: A View from California." *Geological Society of America Special Paper 338*, 55–67.

Brands, H.W., 2002. *The Age of Gold: The California Gold Rush and the New American Dream*. New York: Doubleday.

Briggs, D.E.G., and Fortey R.A., 2005. "Wonderful Strife: Systematics, Stem Groups, and the Phylogenetic Signal of the Cambrian Radiation." *Paleobiology*, v. 31 (no. 2, Supplement), 94–112.

Briggs, D.E.G., Erwin, D.H., and Collier, F.J., 1994. *The Fossils of the Burgess Shale*. Washington, DC and London: Smithsonian Institution Press.

Brock, R.K., ed., 2000. *Emigrant Trails West: A Guide to the California Trail from the Raft River to the Humboldt Sink*. Reno, NV: Trails West.

Bruhn, R.L., DuRoss, C.B., Harris, R.A., and Lund, W.R., 2005. "Neotectonics and Paleoseismology of the Wasatch Fault, Utah." In: Pederson, J., and Dehler, C.M., eds., "Interior Western United States." *Geological Society of America Field Guide 6*, 231–250.

Bryant, E., 1848. *What I Saw in California*. D. Appleton & Co. Reprint, Lincoln: University of Nebraska Press, 1985.

Bryson, B., 2003. *A Short History of Nearly Everything*. New York, NY: Broadway Books.

Burbank, D.W., and Anderson, R.S., 2001. *Tectonic Geomorphology*. Malden, MA, and Oxford, UK: Blackwell Publishing.

Burdick, S., Li, C., Martynov, V., Cox, T., Eakins, J., Mulder, T., Astiz, L., Vernon, F.L., Pavlis, G.L., and van der Hilst, R., 2008. "Upper Mantle Heterogeneity beneath North America from Travel Time Tomography with Global and USArray Transportable Array Data." *Seismological Research Letters*, v. 79, no. 3, 384–392.

Carroll, S., 2005. *Endless Forms Most Beautiful: The New Science of Evo Devo*. New York: W.W. Norton & Company.

Carruthers, M.W., 1999. "Hutton's Unconformity: James Hutton's Ideas about Geology." *Natural History Magazine 108*, no. 5 (June 1999), 86–87.

Caskey, S.J., Bell, J.W., and Slemmons, D.B., 2000. "Historical Surface Faulting and Paleoseismology of the Central Nevada Seismic Belt." In: Lageson, D.R., Peters, S.G., and Lahren, M.M., eds., "Great Basin and Sierra Nevada." *Geological Society of America Field Guide 2*, 23–44.

Chaffin, T., 2002. *Pathfinder: John Charles Frémont and the Course of American Empire*. New York: Farrar, Straus and Giroux.

Chronic, H., 1990. *Roadside Geology of Utah*. Missoula, MT: Mountain Press.

Chronic, H., and Williams, F., 2002. *Roadside Geology of Colorado* (2nd rev. ed.). Missoula, MT: Mountain Press.

Collier, M., 1999. *A Land in Motion: California's San Andreas Fault*. Golden Gate Parks Association (San Francisco) and University of California Press (Berkeley).

Collins, D., 2009. Misadventures in the Burgess Shale. *Nature* (August 2009), v. 460, no. 20, 952–953.

Conway-Morris, S., 1998. *The Crucible of Creation: The Burgess Shale and the Rise of Animals*. Oxford University Press USA.

Corsetti, F.A., and Hagadorn, J.W., 2003. "The Precambrian-Cambrian Transition in the Southern Great Basin, USA." *Sedimentary Record*, v. 1, 4–8.

Courtland, R., 2009. "Earth Calling." *New Scientist*, April 11, 2009, 26–30.

Curran, H., 1982. *Fearful Crossing: The Central Overland Trail through Nevada*. Las Vegas: Nevada Publications.

Dana, J.D., 1873. "On Some Results of the Earth's Contraction from Cooling, Including a Discussion of the Origin of Mountains, and the Nature of the Earth's Interior." *American Journal of Science, 1873*, v. 105, 423–443.

Darwin, C., 1884. *On the Origin of Species by Means of Natural Selection*. New York: D. Appleton.

DeCelles, P.G., 2004. "Late Jurassic to Eocene Evolution of the Cordilleran Thrust Belt and Foreland Basin System, Western U.S.A." *American Journal of Science*, v. 304, 105–168.

DeCourten, F.L., 2003. *The Broken Land: Adventures in Great Basin Geology*. Salt Lake City: University of Utah Press.

Delano, A., 1854. *Life on the Plains and among the Diggings*. Auburn, NY: Miller, Orton & Mulligan. Reprint, Alexandria, VA: Time-Life Books, 1981.

Dickinson, W.R., 2001. "The Coming of Plate Tectonics to the Pacific Rim." In: Oreskes, N., ed., *Plate Tectonics: An Insider's History of the Modern Theory of the Earth*, 264–287. Boulder, CO: Westview.

———, 2002. "The Basin and Range Province as a Composite Extensional Domain." *International Geology Review,* v. 44, 1–38.

———, 2004. "Evolution of the North American Cordillera." *Annual Review of Earth and Planetary Sciences,* v. 32, 13–45.

———, 2006. "Geotectonic evolution of the Great Basin." *Geosphere,* v. 2, 353–368.

———, 2008. "Accretionary Mesozoic–Cenozoic Expansion of the Cordilleran Continental Margin in California and Adjacent Oregon." *Geosphere,* v. 4, no. 2, 329–353.

———, 2009. "Anatomy and Global Context of the North American Cordillera." *Geological Society of America Memoir 204,* 1–29.

Dickinson, W.R., Klute, M.A., Hayes, M.J., Janecke, S.U., Lundin, E.R., McKittrick, M.A., and Olivares, M.D., 1988. "Paleogeographic and Paleotectonic Setting of Laramide Sedimentary Basins in the Central Rocky Mountain Region." *Geological Society of America Bulletin,* v. 100, 1023–1039.

Dickinson, W.R., and Snyder, W.S., 1978, "Plate Tectonics of the Laramide Orogeny." In: Matthews, V., III, ed., "Laramide Folding Associated with Basement Block Faulting in the Western United States." *Geological Society of America Memoir 151,* 355–366.

Dingler, J., Kent, G., Driscoll, N., Babcock, J., Harding, A., Seitz, G., Karlin, B., and Goldman, C., 2009. "A high-Resolution Seismic CHIRP Investigation of Active Normal Faulting across Lake Tahoe Basin, California-Nevada." *Geological Society of America Bulletin,* v. 121, 1089–1107.

Droser, M.L., Jensen, S., and Gehling, J.G., 2002. "Trace Fossils and Substrates of the Terminal Proterozoic–Cambrian transition: Implications for the Record of Early Bilaterians and Sediment Mixing." *Proceedings of the National Academy of Sciences,* v. 99, no. 20, 12752–12756.

DuRoss, C., and McDonald, G., 2005. "The Most Recent Large Earthquake on the Nephi Segment of the Wasatch Fault Zone near Santaquin: Results from 2005 Fault Trenches." *Utah Geological Survey Notes,* January 2007, 1–3.

Elder, W.P., 2001. "Geology of the Golden Gate Headlands." In: Stoffer, P.W., and Gordon, L.C., eds., "Geology and Natural History of the San Francisco Bay Area: A Field-Trip Guidebook." *U.S. Geological Survey Bulletin 2188,* 61–86.

England, P.C., Molnar, P., and Richter, F.M., 2007. "Kelvin, Perry, and the Age of the Earth." *American Scientist,* v. 95, July-August 2007, 342–349.

English, J.M., and Johnston, S.T., 2004. "The Laramide Orogeny: What Were the Driving Forces?" *International Geology Review,* v. 46, 833–838.

English. J.M., Johnston, S.T., and Wang, K., 2003. "Thermal modeling of the Laramide Orogeny: Testing the Flat-Slab Subduction Hypothesis." *Earth and Planetary Sciences Letters 214,* 619–632.

Fey, M., King, R.J., and Lepisto, J., 2002. *Emigrant Shadows: A History and Guide to the California Trail.* Virginia City, NV: Western Trails Research Association.

Flesch, L.M., Holt, W.E., Haines, A.J., and Shen-Tu, B., 2000. "Dynamics of the Pacific-North American Plate Boundary in the Western United States. *Science,* v. 287, 834–836.

Frémont, J.C., 1854. *The Exploring Expedition to the Rocky Mountains, Oregon and California.* Auburn and Buffalo, NY: Miller, Orton and Mulligan.

Fromm, R., Zandt, G., and Beck, S.L., 2004. "Crustal thickness beneath the Andes and Sierras Pampeanas at 30S Inferred from Pn Apparent Phase Velocities." *Geophysical Research Letters*, v. 31, L06625.

Gaines, R.R., Kennedy, M.J., and Droser, M.L., 2005. "A New Hypothesis for Organic Preservation of Burgess Shale Taxa in the Middle Cambrian Wheeler Formation, House Range, Utah." *Palaeogeography, Palaeoclimatology, Palaeoecology*, v. 220, 193–205.

Gilette, D.D., 1996. "Utah's Wildlife in the Ice Age." *Utah Geological Survey Notes*, v. 28, no. 3 (May 1996), available online at: http://geology.utah.gov/utahgeo/dinofossil/iceage/icewildlife.htm.

Godsey, H.S., Atwood, G., Lips, E., Miller, D.M., Milligan, M., and Oviatt, C.G., 2005. "Don R. Currey Memorial Field Trip to the Shores of Pleistocene Lake Bonneville." In: Pederson, J., and Dehler, C.M., eds., "Interior Western United States." *Geological Society of America Field Guide 6*, 419–448.

Gould, S.J, 1987. *Time's Arrow, Time's Cycle*. Cambridge, MA: Harvard University Press.

———, 1990. *Wonderful Life: The Burgess Shale and the Nature of History*. New York: W.W. Norton.

———, 1996. *Full House*. New York: Harmony Books.

Graham, B., 2003. "Frémont and the Determination of Elevations." Online article at: http://www.longcamp.com/hypsometry.html.

Graydon, C.K., 1986. "Trail of the First Wagons over the High Sierra." *Overland Journal*, v. 4, no. 2, 4–15.

Gregory, K.M., and Chase, C.G., 1992. "Tectonic Significance of Paleobotanically Estimated Climate and Altitude of the Late Eocene Erosion Surface, Colorado." *Geology*, v. 20, 581–585.

Grieve, D.D., Hough, A.S., and Littlejohn, D., 1974. *Hard-Rock Gold Mining*. DVD produced and distributed by California State Parks, Sacramento, CA.

Gwynn, J.W., 1996. "Commonly Asked Questions about Utah's Great Salt Lake and Ancient Lake Bonneville." *Utah Geological Survey Public Information Series 39*.

Hagadorn, J.W., 2002a. "Chengjiang: Early Record of the Cambrian Explosion." In: Bottjer, D.J., et al., eds., *Exceptional Fossil Preservation: A Unique View on the Evolution of Marine Life*, 35–60. New York: Columbia University Press.

———, 2002b, "Burgess Shale: Cambrian Explosion in Full Bloom." In: Bottjer, D.J., et al., eds., *Exceptional Fossil Preservation: A Unique View on the Evolution of Marine Life*, 61–89. New York: Columbia University Press.

———, 2002c, "Burgess Shale-Type Localities: The global Picture." In: Bottjer, D.J., et al., eds., *Exceptional Fossil Preservation: A Unique View on the Evolution of Marine Life*, 91–116. New York: Columbia University Press.

———, 2009. "Taking a Bite Out of *Anomalocaris*." In: Smith, M.R., O'Brien, L.J., and Caron, J.B., *International Conference on the Cambrian Explosion, Abstract Volume*. Toronto, Ontario, Canada: Burgess Shale Consortium.

Hagadorn, J.W., and Waggoner, B.M., 2000, "Ediacaran Fossils from the Southwestern Great Basin, United States." *Journal of Paleontology*, v. 74, 349–359.

Hagadorn, J.W., Fedo, C.W., and Waggoner, B.M., 2000, "Early Cambrian Ediacaran-Type Fossils from California." *Journal of Paleontology*, v. 74, 731–740.

Hamblin, W.K., and Christiansen, E.H., 2004. *Earth's Dynamic Systems* (10th ed.). Upper Saddle River, NJ: Pearson Education.

Hanson, R.E., Girty, G.H., Harwood, D.S., and Schweickert, R.A., 2000. "Paleozoic Subduction Complex and Paleozoic-Mesozoic Island-Arc Volcano-Plutonic Assemblages in the Northern Sierra Terrane." In: Lageson, D.R., Peters, S.G., and Lahren, M.M., eds., "Great Basin and Sierra Nevada." *Geological Society of America Field Guide Series,* v. 2, 255–277.

Harden, D.R., 2004. *California Geology* (2nd ed.). Upper Saddle River, NJ: Pearson Prentice Hall.

Hastings, L.W., 1845. *The Emigrant's Guide to Oregon and California.* Cincinnati: George Conclin; stereotyped by Shepard and Company, 1845. Online version at: http://www.scienceviews.com/historical/emigrantguide.html.

Hayes, D., 2007. *Historical Atlas of California.* Berkeley, CA: University of California Press.

Henry, C.D., 2008. "Ash-Flow tuffs and Paleovalleys in Northeastern Nevada: Implications for Eocene Paleogeography and Extension in the Sevier Hinterland, Northern Great Basin." *Geosphere,* v. 4, no. 1, 1–35.

Henry, C.D., and Ressel, M.W., 2000. "Interrelation of Eocene Magmatism, Extension, and Carlin-Type Gold Deposits in Northeastern Nevada." In: Lageson, D.R., Peters, S.G., and Lahren, M.M., eds., "Great Basin and Sierra Nevada, Boulder, Colorado." *Geological Society of America Field Guide 2,* 165–187.

Hill, M., 2002. *Gold: The California Story.* Berkeley, CA: University of California Press.

———, 2006. *Geology of the Sierra Nevada, California Natural History Guide Series No. 80.* Berkeley, CA: University of California Press.

Hintze, L.F., and Davis, F.D., 2003. *Geology of Millard County, Utah.* Bulletin 133, Utah Geological Survey, Salt Lake City.

Hintze, L.F., and Robison, R.A., 1987. "The House Range, Western Utah: Cambrian Mecca." *Geological Society of America Centennial Field Guide*—Rocky Mountain Section. Boulder, CO: Geological Society of America, 257–260.

Hofstra, A.H., and Wallace, A.R., 2006. "Metallogeny of the Great Basin: Crustal Evolution, Fluid Flow, and Ore Deposits." *U.S. Geological Survey Open-File Report* 2006-1280.

Holliday, J.S., 1981. *The World Rushed In: The California Gold Rush Experience.* New York: Simon & Schuster.

———, 1999. *Rush to Riches: Gold Fever and the Making of California.* Berkeley, CA: Oakland Museum of California and the University of California Press.

Horton, T.W., and Chamberlain, C.P., 2006. "Stable Isotopic Evidence for Neogene Surface Downdrop in the Central Basin and Range Province." *Geological Society of America Bulletin,* v. 118, no. 3-4, 475–490.

Hough, A.S., Grieve, D.D., Ward, T., and Littlejohn, D., 1973. *Hydraulic Gold Mining.* DVD produced and distributed by California State Parks, Sacramento, CA.

Hough, S.E., 2004. *Finding Fault in California: An Earthquake Tourist's Guide.* Missoula, MT: Mountain Press.

Houghton, E.D., 1911. *The Expedition of the Donner Party and its Tragic Fate.* Chicago: A.C. McClurg and Co.; reprinted by the University of Nebraska Press (Lincoln, NE) 1997.

Humphreys, E., 2009. "Relation of Flat Subduction to Magmatism and Deformation in the Western United States." *Geological Society of America Memoir 204,* 85–98.

Hutton, J., 1788. "Theory of the Earth; Or, An Investigation of the Laws Observable in the Composition, Dissolution, and Restoration of Land upon the Globe." *Transactions of the Royal Society of Edinburgh,* v. 1, 209–304.

Ingersoll, R.V., 2000. "Models for Origin and Emplacement of Jurassic Ophiolites of Northern California." In: Dilek, Y., Moores, E.M., Elthon, D., and Nicolas, A., eds., "Ophiolites and Oceanic Crust: New Insights from Field Studies and the Ocean Drilling Program." *Geological Society of America Special Paper 349,* 395–402.

James, R.M., 1998. *The Roar and the Silence: A History of Virginia City and the Comstock Lode.* Reno, Nevada: University of Nevada Press.

Jensen, S., Droser, M.L., and Gehling, J.G., 2004. "Trace Fossil Preservation and the Early Evolution of Animals." *Palaeogeography, Palaeoclimatology, Palaeoecology,* v. 220, 19–29.

Karl, H.A., Chin, J.L., Ueber, E., Stauffer, P.H., and Hendley, J.W., II, 2001. "Beyond the Golden Gate—Oceanography, Geology, Biology, and Environmental Issues in the Gulf of the Farallons." *U.S. Geological Survey Circular 1198.*

Karlstrom, K.E., Crow, R., Crossey, L.J., Coblentz, D., and Van Wijk, J.W., 2008. "Model for Tectonically Driven Incision of the <6 Ma Grand Canyon." *Geology,* v. 36, 835–838.

Kent, G.M, et al., 2005. "60 K.Y. Record of Extension across the Western Boundary of the Basin and Range Province: Estimate of Slip Rates from Offset Shoreline Terraces and a Catastrophic Slide beneath Lake Tahoe." *Geology,* v. 33, no. 5, 365–368.

Konigsmark, T., 2002. *Geologic Trips: Sierra Nevada.* Gualala, CA: GeoPress.

Lageson, D.R., and Spearing, D.R., 1991. *Roadside Geology of Wyoming* (2nd rev. ed.). Missoula, MT: Mountain Press.

Levy, J., 1990. *They Saw the Elephant: Women in the California Gold Rush.* Hamden, CT: Shoe String Press.

Love, D.L., and Christiansen, A.C., 1985. *Geologic Map of Wyoming, scale 1:500,000.* Denver: Department of the Interior, U.S. Geological Survey.

Lubetkin, L.K.C., and Clark, M.M., 1987. "Late Quaternary Fault Scarp at Lone Pine, California: Location of Oblique Slip during the 1872 Earthquake and Earlier Earthquakes." Geological Society of America Centennial Field Guide—Cordilleran Section. Boulder, CO: *Geological Society of America,* 151–156.

Liu, L., Gurnis, M., Seton, M., Saleeby, J., Müller, R.D., and Jackson, J.M., 2010. "The Role of Oceanic Plateau Subduction in the Laramide Orogeny." *Nature Geoscience,* v. 3, 353–357.

Macdougall, D., 2008. *Nature's Clocks: How Scientists Measure the Age of Almost Everything.* Berkeley, CA: University of California Press.

Maher, H.D., Jr., Englemann, G.F., and Shuster, R.D., 2003. *Roadside Geology of Nebraska.* Missoula, MT: Mountain Press.

Marshak, S., 2008. *Earth: Portrait of a Planet* (3rd ed.). New York: W.W. Norton.

Marshall, C.R., 2006. "Explaining the Cambrian 'Explosion' of animals." *Annual Review of Earth and Planetary Sciences,* v. 34, 355–384.

Martin, M.W., Grazhdankin, D.V., Bowring, S.A., Evans, D.A.D., Fedonkin, M.A., Kirsch-vink, J.L., 2000. "Age of Neoproterozoic Bilaterian Body and Trace Fossils, White Sea, Russia: Implications for Metazoan Evolution." *Science*, v. 288, 841–845.

Marvine, A.R., 1874. "Report for the Year 1873." In: "7th Annual Report of the U.S. Geological and Geographical Survey of the Territories (Hayden Survey)," Washington, DC, 83–192.

McCall, A.J., 1882. *The Great California Trail in 1849: Wayside Notes of an Argonaut.* Bath, NY: Steuben Courier. Reprinted from the *Steuben Courier.*

McCalpin, J.P., and Nishenko, S.P., 1996. "Holocene Paleoseismicity, Temporal Clustering, and Probabilities of Future Large (M >7) Earthquakes on the Wasatch Fault Zone, Utah." *Journal of Geophysical Research*, v. 101, 6233–6253.

McLynn, F., 2002. *Wagons West: The Epic Story of America's Overland Trails.* New York: Grove Press.

McMenamin, M.A.S., 1998. *The Garden of Ediacara.* New York: Columbia University Press.

McMillan, M.E., Angevine, C.L., and Heller, P.L., 2002. "Postdepositional Tilt of the Miocene-Pliocene Ogallala Group on the Western Great Plains: Evidence of Late Ceno-zoic Uplift of the Rocky Mountains." *Geology*, v. 30, 63–66.

McMillan, M.E., Heller, P.L., and Wing, S.L., 2006. "History and Causes of Post-Laramide Relief in the Rocky Mountain Orogenic Plateau." *Geological Society of America Bulletin*, v. 118, 393–405.

McPhee, J., 1998. *Annals of the Former World.* New York: Farrar, Straus and Giroux.

Mears, B., Jr., 1993. "Geomorphic History of Wyoming and High-Level Erosion Surfaces." In: Snoke, A.W., Steidtmann, J.R., and Roberts, S.M., eds., "Geology of Wyoming, Lara-mie, WY." *Geological Survey of Wyoming Memoir 5*, 608–626.

Mears, B., Jr., Eckerle, W.P., Gilmer, D.R., Gubbles, T.L., Huckleberry, G.A., Marriot, H.J., Schmidt, K.J., and Yose, L.A., 1986. "A Geologic Tour of Wyoming from Laramie to Lander, Jackson and Rock Springs." *Geological Survey of Wyoming Public Information Circular No. 27.*

Meldahl, K.H., 2007. *Hard Road West: History and Geology along the Gold Rush Trail.* Chi-cago: University of Chicago Press.

Midpeninsula Regional Open Space District. *San Andreas Fault Trail Guide: Los Trancos Open Space Preserve.* Online at http://www.openspace.org/activities/downloads/San_An-dreas_Fault_Trail.pdf.

Miller, D.E., 1958. "The Donner Road through the Great Salt Lake Desert." *Pacific Historical Review*, v. 27, no.1, 39–44.

Moeller, B., and Moeller, J., 2001. *The Oregon Trail: A Photographic Journey.* Missoula, MT: Mountain Press.

Moores, E.M., 1970. "Ultramafics and Orogeny, with Models of the U.S. Cordillera and the Tethys." *Nature*, v. 228, 837–842.

Moores, E.M., 2003. "A Personal History of the Ophiolite Concept." In: Dilek, Y., and New-comb, S., eds., "Ophiolite Concept and the Evolution of Geologic Thought." *Geological Society of America Special Paper 373*, 17–29.

Moores, E.M., Dilek, Y., and Wakabayashi, J., 1999. "California Terranes." In: "Classic Cor-dilleran Concepts: A View from California." Moores, E.M., Sloan, D., and Stout, D.L., eds., *Geological Society of America Special Paper 338*, 221–234.

Moores, E.M., Wakabayashi, J., and Unruh, J.R., 2002. Crustal-Scale Cross-Section of the U.S. Cordillera, California and beyond, Its Tectonic Significance, and Speculations on the Andean Orogeny." *International Geology Review*, v. 44, 479–500.

Moores, E.M., Wakabayashi J., Unruh J., and Waechter S., 2006. "A Transect Spanning 500 Million Years of Active Plate Margin History: Outline and Field Trip Guide." *Geological Society of America Field Guide 7*: 1906 San Francisco Earthquake, 373–413.

Moreno, R., 2000. *Roadside History of Nevada*. Missoula, MT: Mountain Press.

Muir, J., 1901. *Our National Parks*. Boston: Houghton-Mifflin.

————, 1914. *The Yosemite*. New York: Century Company.

Mulch, A., Graham, S.A., and Chamberlain, C.P., 2006. "Hydrogen Isotopes in Eocene River Gravels and Paleoelevation of the Sierra Nevada." *Science*, v. 313, 87–89.

Muntean, J., 2006. "The Rush to Uncover Gold's Origins." *Geotimes*, April 2006, 24–27.

————, 2008a, "Metals." In: "The Nevada Mineral Industry 2007." *Nevada Bureau of Mines and Geology Special Publication MI-2007*, 27–87.

————, 2008b. "Major Precious Metal Deposits." In: "The Nevada Mineral Industry 2007." *Nevada Bureau of Mines and Geology Special Publication MI-2007*, 88–116.

Niemi, N.A., Wernicke, B.P., Friedrich, A.M., Simmons, M., Bennett, R.A., and Davis, J.L., 2004. "BARGEN Continuous GPS Data across the Eastern Basin and Range Province, and Implications for Fault System Dynamics." *Geophysical Journal International*, v. 159, no. 3, 842–862.

Niemi, T.M., Hall, N.T., and Dahne, A., 2006. "The 1906 Earthquake Rupture Trace of the San Andreas Fault North of San Francisco, with Stops at Points of Geotechnical Interest." *Geological Society of America Field Guide 7*: 1906 San Francisco Earthquake, 157–176.

O'Connor, J.E., 1993. "Hydrology, Hydraulics, and Geomorphology of the Bonneville Flood." *Geological Society of America Special Paper 274*.

O'Connor, J.E., and Costa, J.E., 2004. "The World's Largest Floods, Past and Present— Their Causes and Magnitudes." *U.S. Geological Survey Circular 1254*.

O'Connor, J.E., Grant, G.E., and Costa, J.E., 2002. "The Geology and Geography of Floods." In: House, P.K., Webb, R.H., Baker, V.R., and Levish, D.R., eds., "Ancient Floods, Modern Hazards: Principles and Applications of Paleoflood Hydrology." *Water Science and Application*, v. 5. American Geophysical Union, Washington, D.C., 359–385.

O'Neil, J., Carlson, R.W., Francis, D., and Stevenson, R.K., 2008. "Neodymium-142 Evidence for Hadean Mafic Crust." *Science*, v. 321. no. 5897, 1828–1831.

Oreskes, N., 2001. "From Continental Drift to Plate Tectonics." In: Oreskes, N., ed., *Plate Tectonics—An Insider's History of the Modern Theory of the Earth*, 3–27. Cambridge, MA: Westview Press.

Orndorff, R. L., Wieder, R.W., and Filkorn, H.F., 2001. *Geology Underfoot in Central Nevada*. Missoula, MT: Mountain Press.

Oskin, M., Stock, J.M., and Martín-Barajas, A., 2001. "Rapid Localization of Pacific–North America Plate Motion in the Gulf of California." *Geology*, v. 29, 459–462.

Oviatt, C.G., 1997. "Lake Bonneville Fluctuations and Global Climate Change." *Geology*, v. 25, 155–158.

Parke, C.R., 1989. *Dreams to Dust: A Diary of the California Gold Rush, 1849–1850*. Davis, J.E., ed. Lincoln, NE: University of Nebraska Press.

Pelletier, J.D., 2009. "The Impact of Snowmelt on the Late Cenozoic Landscape of the Southern Rocky Mountains, USA." *GSA Today*, v. 17, no. 7, 4–11.

Plataforma., 2009. "Prehistoric Bears Ate Everything and Anything, Just Like Modern Cousins." *ScienceDaily* April 13, 2009. Accessed June 9, 2010 from http://www.science daily.com/releases/2009/04/090408170815.htm.

Potter, D.M., ed., 1945. *Trail to California: The Overland Journal of Vincent Geiger and Wakeman Bryarly*. New Haven, CT: Yale University Press.

Poulsen, R.C., 1977. " 'This Is the Place': Myth and Mormondom." *Western Folklore*, v. 36, no. 3, 246–252.

Powell, J.W., 1875. *Exploration of the Colorado River of the West and Its Tributaries. Explored in 1869, 1870, 1871, and 1872, Under the Direction of the Secretary of the Smithsonian Institution*. Washington, DC: U.S. Government Printing Office.

———, 1895. *Canyons of the Colorado*. Meadville, PA: Flood & Vincent.

Price, J.G., 2007. "Overview." In: "The Nevada Mineral Industry 2007." *Nevada Bureau of Mines and Geology Special Publication MI-2007*, 3–26.

Ramelli, A.R., Bell, J.W., dePolo, C.M., and Yount, J.C., 1999. "Large-Magnitude, Late Holocene Earthquakes on the Genoa Fault, West-Central Nevada and Eastern California." *Bulletin of the Seismological Society of America*, v. 89, 1458–1472.

Ramos, V.A., 2008. "Field Trip Guide: Evolution of the Pampean Flat-Slab Region over the Shallowly Subducting Nazca Plate." *The Geological Society of America Field Guide 13*, 77–116.

———, 2009. "Anatomy and Global Context of the Andes: Main Geologic Features and the Andean Orogenic Cycle." *Geological Society of America Memoir 204*, 31–65.

Ramos, V.A., Cristallini, E.O., and Pérez, D.J., 2002. "The Pampean Flat-Slab of the Central Andes." *Journal of South American Earth Sciences*, v. 15, no. 1, 59–78.

Raney, W., 2005. *Carving Grand Canyon*. Grand Canyon, AZ: Grand Canyon Association.

Rarick, E., 2008. *Desperate Passage: The Donner Party's Perilous Journey West*. Oxford University Press, USA.

Richet, P., 2007. A Natural History of Time. Translated by John Venerella. Chicago, IL: University of Chicago Press.

Saleeby, J., 2003. Segmentation of the Laramide Slab: Evidence from the Southern Sierra Nevada." *Geological Society of America Bulletin 115*, 655–668.

Scamehorn, H.L. 1965. The Buckeye Rovers in the Gold Rush: An Edition of Two Diaries. Athens, OH: Ohio University Press.

Schweickert, R.A., Lahren, M.M., Karlin, R., and Howle, J., 2000. "Lake Tahoe Active Faults, Landslides, and Tsunamis." In: Lageson, D.R., Peters, S.G., and Lahren, M.M., eds., "Great Basin and Sierra Nevada." *Geological Society of America Field Guide 2*, 1–21.

Sharp, R.P., and Glazner, A.F., 1997. *Geology Underfoot in Death Valley and Owens Valley*. Missoula, MT: Mountain Press, 195–201.

Sieh, K.E., and Wallace, R.E., 1987. "The San Andreas Fault at Wallace Creek, San Luis Obispo County, California." In: Hill, M.L., ed., *Geological Society of America Centennial Field Guide—Cordilleran Section*. Boulder, CO: Geological Society of America, 233–238.

Sieh, K.E., and Jahns, R.H., 1984. "Holocene Activity of the San Andreas Fault at Wallace Creek, California." *Geological Society of America Bulletin*, v. 95, 883–896.

Sigloch, K., McQuarrie, N., and Nolet, G., 2008. "Two-Stage Subduction History under North America Inferred from Multiple-Frequency Tomography." *Nature Geoscience*, v. 1, 458–462.

Silva, M., 1986. *Placer Gold Recovery Methods.* Sacramento: California Department of Conservation, Division of Mines and Geology Special Publication 87.

Slemmons, D.B., and Bell, J.W., 1987. "1954 Fairview Peak Earthquake Area, Nevada." *Geological Society of America Centennial Field Guide—Cordilleran Section,* 73–76. Boulder, CO: Geological Society of America.

Smith, J.F., and Ketner, K.B., 1975. "Stratigraphy of Paleozoic Rocks in the Carlin-Pinon Range Area, Nevada." *U.S. Geological Survey Professional Paper 867-A.*

Snoke, A.W., 1993. "Geologic History of Wyoming within the Tectonic Framework of the North American Cordillera." In: Snoke, A.W., Steidtmann, J.R., and Roberts, S.M., eds., *Geology of Wyoming,* 1993. Laramie, WY: Geological Survey of Wyoming Memoir no. 5, 2–56.

Solomon, B.J., Ashland, F.X., Giraud, R.E., Hylland, M.D., Black, B.D., Ford, R.L., Hernandez, M.W., and Hart, D.H., 2005. "Geologic Hazards of the Wasatch Front, Utah." In: Pederson, J., and Dehler, C.M., eds., *Interior Western United States: Geological Society of America Field Guide 6, Boulder, CO: Geological Society of America,* 505–524.

Southern California Earthquake Center. *Wallace Creek Interpretive Trail Guide: A Geologic Guide to the San Andreas Fault at Wallace Creek.* Southern California Earthquake Center online at: http://www.scec.org/wallacecreek/pdf/trailguide.pdf.

Stegener, W., 1954. *Beyond the Hundredth Meridian: John Wesley Powell and the Opening of the West.* Cambridge, MA: Houghton Mifflin.

Stewart, J.H., and Carlson, J.E., 1978. *Geologic Map of Nevada, Scale 1:500,000 and Accompanying Text.* Denver, CO: U.S. Geological Survey.

Stock, G.M., Anderson, R.S., and Finkel, R.C., 2004. "Pace of Landscape Evolution in the Sierra Nevada, California, Revealed by Cosmogenic Dating of Cave Sediments." *Geology,* v. 32, 193–196.

Stoffer, P.W., 2006. "Where Is the San Andreas Fault? A Guidebook to Tracing the Fault on Public Lands in the San Francisco Bay Region." *U.S. Geological Survey General Interest Publication 16,* http://pubs.usgs.gov/gip/2006/16/.

Suess, E., 1885, *Das Antlitz der Erde* [The Face of the Earth]. Published in four volumes between 1888 and 1909. F. Tempsky, Prag. 2776 pages.

Thatcher, W., Foulger, G.R., Julian, B.R., Svarc, J., Quilty, E., and Bawden, G.W., 1999. "Present-Day Deformation across the Basin and Range Province, Western United States." *Science,* v. 283. no. 5408, 1714–1718.

Thompson, T.B., Teal, L., and Meeuwig, R.O., 2002. *Gold Deposits of the Carlin Trend.* Nevada Bureau of Mines and Geology Bulletin 111. Reno, NV: University of Nevada Press.

Tingley, J.V., and Pizarro, K.A., 2000. *Traveling America's Loneliest Road—A Geologic and Natural History Tour through Nevada along U.S. Highway 50.* Reno, NV: University of Nevada Press.

Trexler, J.H., Jr., Cashman, P.H., Snyder, W.S., and Davidov, V.I., 2004. "Late Paleozoic Tectonism in Nevada: Timing, Kinematics, and Tectonic Significance." *Geological Society of America Bulletin,* v. 116, 525–538.

Trimble, D.E., 1980. *The Geologic Story of the Great Plains*. U.S. Geological Survey Bulletin 1493. Washington, DC: U.S. Government Printing Office.

Twain, M., 1872. *Roughing It*. American Publishing Company, reprinted 1981 by Penguin Books.

Tweeto, O., 1979, *Geologic Map of Colorado, Scale 1:500,000*. Denver, CO: Department of the Interior, U.S. Geological Survey.

University of Arizona, 1998. "New Technique for Analyzing DNA in Fossil Dung Could Help Scientists Sort the Details of Megafauna Extinction." *ScienceDaily* July 21, 1998. Online at http://www.sciencedaily.com /releases/1998/07/980721081204.htm.

Unruh, J., Humphrey, J., and Barron, A., 2003. "Transtensional Model for the Sierra Nevada Frontal Fault System, Eastern California." *Geology* v. 33, no. 4, 327–330.

U.S. Geological Survey Earthquake Hazards Program—Seismicity of the Western United States: 1990–2000, http://earthquake.usgs.gov/regional/states/seismicity/us_west_seismicity.php

U.S. Geological Survey Earthquake Hazards Program—Seismicity of the Western United States: Historic Earthquakes of California, http://earthquake.usgs.gov/regional/states/california/history.php.

U.S. Geological Survey Earthquake Hazards Program—Seismicity of the Western United States: Historic Earthquakes of Nevada, http://earthquake.usgs.gov/regional/states/nevada/history.php.

U.S. Geological Survey Utah Water Science Center. "Brine Shrimp and Ecology of Great Salt Lake." http://ut.water.usgs.gov/greatsaltlake/shrimp/.

U.S. Geological Survey Utah Water Science Center. "Great Salt Lake—Salinity and Water Quality." http://ut.water.usgs.gov/greatsaltlake/salinity/.

Utah Geological Survey 1996. *The Wasatch Fault*. Public Information Series 40.

Valdiya, K.S., 1996. "Antecedent Rivers: Ganga Is Older Than Himalaya." *Resonance*, v. 1, no. 8, 55–63.

———, 1998. *Dynamic Himalaya*. Jakkur, India: Jawaharlal Nehru Centre for Advanced Scientific Research.

van Hunen, J., van den Berg, A.P., and Vlaar, N.J., 2002. "On the Role of Subducting Oceanic Plateaus in the Development of Shallow Flat Subduction." Tectonophysics, v. 352, 317–333.

Wahrhaftig, C., and Murchey, B., 1987. "Marin Headlands, California: 100-Million-Year Record of Sea Floor Transport and Accretion." In: Hill, M.L., ed., *Geological Society of America Centennial Field Guide—Cordilleran Section*. Boulder, CO: Geological Society of America, 263–268.

Wakabayashi, J., and Sawyer, T.L., 2001. "Stream Incision, Tectonics, Uplift, and Evolution of Topography of the Sierra Nevada." *Journal of Geology*, v. 109, 539–562.

Wakabayashi, J., and Stock, G.M., 2003. "Overview of the Cenozoic Geologic History of the Sierra Nevada." In: Stock, G.M., ed., *Tectonics, Climate Change, and Landscape Evolution in the Southern Sierra Nevada, California: Pacific Cell Friends of the Pleistocene Field Trip Guidebook*. Santa Cruz, CA: University of California at Santa Cruz, 31–40.

Wallace, R.E., 1987. "Fault Scarps Formed During the Earthquakes of October 2, 1915, Pleasant Valley, Nevada." *Geological Society of America Centennial Field Guide—Cordilleran Section*. Boulder, CO: Geological Society of America, 83–84.

Wilde, S.A., Valley, J.W., Peck, W.H., and Graham, C.M., 2001. "Evidence from Detrital Zircons for the Existence of Continental Crust and Oceans on the Earth 4.4 Gyr Ago." *Nature*, v. 409, 175–178.

Wilson, M.L., and Dennis, R.D., 2006. *An Integrated Petroleum Evaluation of Northeastern Nevada*. (The following online section of this report covers the Bates Mountain Tuff: http://westerncordillera.com/bates_mountain_tuff.htm.)

Wolfe, J.A., Forest, C.E., and Molnar, P., 1998. "Paleobotanical Evidence of Eocene and Oligocene Paleoatitudes in Midlatitude Western North America." *Geological Society of America Bulletin*, v. 110, no. 5, 664–678.

Wolfe, J.A., Schorn, H.E., Forest, C.E., and Molnar, P., 1997. "Paleobotanical Evidence for High Altitudes in Nevada during the Miocene." *Science*, v. 276, 1672–1675.

Wyckoff, R.M., 1999. *Hydraulicking North Bloomfield and the Malakoff Diggins State Historic Park*. Nevada City, CA: Robert M. Wyckoff.

Zandt, G., and Humphreys, G., 2008. "Toroidal Mantle Flow through the Western U.S. Slab Window." *Geology*, v. 36, 295–298.

FIGURE SOURCES AND CREDITS

All shaded relief bases on shaded relief maps are from the National Elevation Dataset of the U.S. Geological Survey's Earth Resources Observation and Science (EROS) Center or from NASA's Jet Propulsion Laboratory. All photographs and illustrations not credited below are original work of the author.

Frontispiece   The Earth's Tectonic Plates map modified from NASA's Digital Tectonic Activity Map of the Earth, http://denali.gsfc.nasa.gov/dtam/.

Figure 1.1   Earth cross-section diagram from Meldahl 2007, fig. 7.2, p. 113, reproduced with permission of the University of Chicago Press.

Figure 1.2 (top)   Point Bonita aerial photo from the National Oceanographic and Atmospheric Administration's Office of National Marine Sanctuaries, derived from an online article dated November 15, 2007, titled "NOAA, Partners Remain Active in Spill Response," http://www.noaanews.noaa.gov/stories2007/20071115_oilspill.html.

Figure 1.3   Tectonic system diagram modified from digital art supplied to the public domain by the U.S. Geological Survey.

Figure 1.4   Map of terranes of western North America based on Marshak 2008, fig. 11.34, p. 387.

Figure 1.6 (top)   Oblique virtual image of the accretionary wedge forming today off the coast of Oregon reproduced with permission of Lincoln Pratson, from fig. 1 in Pratson, L.F. and W. Haxby, 1996, What is the slope of the U.S. continental slope? *Geology* 24, pp. 3–6.

Figure 1.6 (bottom)   Block diagram of the Oregon accretionary wedge based on fig. 21.10 in Hamblin and Christiansen, 2004, p. 607.

Figure 2.2   California's growth from accreted terranes based on DeCourten 2003, fig. 2.4, p. 28.

Figure 2.3   Sierra Nevada terranes based on Harden 2004, fig. 8-25, p. 183.

Figure 2.7   Sierra lode gold districts based on Böhlke 1999, fig. 1B.

Figure 3.2   Malakoff Diggins photo: Hydraulic mining at Malakoff Diggings (circa 1876), located in the South Yuba River watershed, California. Hearst Collection of Mining Views by Carleton E. Watkins, reproduced with permission of the Bancroft Library, University of California, Berkeley.

Figure 3.3   California gold production graph modified from Meldahl 2007, fig. 13.6, p. 267, originally based on Harden 2004, fig. 8-37, p. 198, with additional data supplied by the California Geological Survey.

Figure 3.4   Fossil rivers map based on Harden 2004, fig. 8.31, p. 192.

Figure 3.5   Formation of the Sierra Nevada and uplift of fossil riverbeds based on text in Mulch, Graham, and Chamberlain 2006 and in DeCelles 2004, as well as on personal communication with Page Chamberlain.

Figure 4.5   Donner Pass ascent: Painting by Harold von Schmidt, "Rough Going Over the Sierras." Date unknown. From a digital image supplied to the author by Encore Editions.

Figure 4.6   Ascent of Roller Pass: Based on Stewart, G. R. 1962, *The California Trail: An Epic with Many Heroes* (Lincoln: University of Nebraska Press), p. 117.

Figure 5.1 (bottom)   Aerial photograph of the San Andreas fault in the Carrizo Plain reproduced with permission of Michael Collier.

Figure 5.2   Maps of earthquakes and plate edges based on the U.S. Geological Survey National Earthquake Information Center, adopted from student laboratory exercises developed by Tanya Atwater, University of California at Santa Barbara.

Figure 5.4 (top)   Aerial photograph of the San Andreas fault at Wallace Creek reproduced with permission of Michael Collier.

Figure 5.4 (bottom)   Time diagrams of the San Andreas fault at Wallace Creek adopted from diagrams by Aron Meltzner in the Southern California Earthquake Center's Wallace Creek Interpretive Trail Guide, http://www.scec.org/wallacecreek/pdf/trailguide.pdf.

Figure 5.6   Map of differential movement of regions in the western United States based on Flesch et al. 2000, fig. 1, p. 834.

Figure 6.1   Rio Buenaventura map by John Melish 1816: Library of Congress: G3700 1816.M4f Mel.

Figure 6.2   Rio Buenaventura map by Finley 1826: from Bancroft, H. H., 1889, History of Utah, 1540–1886. San Francisco: The History Company, p. 28.

Figure 6.3   Virtual satellite-view image looking obliquely across the Great Basin reproduced with permission of Dr. William Bowen, California Geographical Survey.

Figure 6.4   Great Basin diagram modified from Meldahl 2007, fig. 10.9, p. 202, reproduced with permission of the University of Chicago Press, based on DeCourten 2003, fig. 8.17, p. 185.

Figure 6.5 (top)   Carson Range cross-section diagram based on Schweickert et al. 2000, fig. 7, p. 10.

Figure 6.5 (bottom)   Photograph of the Genoa fault from Meldahl 2007, fig. 12.9, p. 247, reproduced with permission of the University of Chicago Press.

Figure 6.6   Great Basin tectonic history time diagrams created in consultation with Pete DeCelles, Department of Geosciences, University of Arizona, and adopted from Meldahl 2007, figs. 7.6, 7.11, and 10.8 (pp. 123, 135, and 200, respectively); reproduced with permission of the University of Chicago Press.

Figure 6.7   Mountain formation diagrams modified from digital art supplied to the public domain by the U.S. Geological Survey.

Figure 6.8   Time maps of Basin and Range and San Andreas fault development based on Marshak 2008, fig. 13.32, p. 475, derived originally from Atwater 1970.

Figure 7.2 (top)   Map of seismic belts based on U.S. Geological Survey Earthquake Hazards Program, Seismicity of the Western United States.

Figure 7.2 (bottom)   Graph of extension rates across the Basin and Range and San Andreas system based on Bennett, Davis, and Wernicke 1999, Thatcher et al. 1999, and Niemi et al. 2004.

Figure 7.3   Tobin Range 1915 fault scarp photograph by Robert E. Wallace, U.S. Geological Survey Photographic Library, originally published in Earthquake Information Bulletin, v. 6, no. 6, p. 29.

Figure 7.4 (top)   1954 photograph of Dixie Valley fault scarp from the Karl V. Steinbrugge Collection, University of California, Berkeley.

Figure 7.4 (bottom)   Seismic reflection image across the Dixie Valley fault from Abbott, R. E., J. N. Louie, S. J. Caskey, S. Pullammanappallil, 2001, Geophysical confirmation of low-angle normal slip on the historically active Dixie Valley fault, Nevada, *Journal of Geophysical Research*, 106, B3, 4169–4181, fig. 4. Reproduced in modified form by permission of the American Geophysical Union.

Figure 7.5   Comparison diagrams of slow-rising versus fast-rising mountain ranges based on Burbank and Anderson 2001, fig. 10.1, p. 202.

Figure 7.6 (bottom)   Aerial photograph of the Wasatch front and Wasatch fault reproduced with permission of the Utah Geological Survey (photograph by Rod Millar).

Figure 8.3   Tectonic time diagrams related to metal ore formation modified from Meldahl 2007, fig. 10.8, p. 200; reproduced with permission of the University of Chicago Press.

Figure 8.4   Diagram of caldera formation modified from DeCourten 2003, fig. 8.10, p. 178; reproduced with permission of the University of Utah Press.

Figure 8.5   Photograph of the eruption of Sarychev Volcano in the Kuril Islands taken from the International Space Station on June 12, 2009. Credit: NASA.

Figure 8.6 (top)   Photograph of the city of Martinique after Mt. Pelee eruption: Library of Congress Prints and Photographs.

Figure 8.6 (bottom)   Photograph of Pompeii dog reproduced with permission of Samantha Tengelitsch.

Figure 9.1   Photograph of melon gravels along Snake River reproduced with permission of David Diller.

Figure 9.2   Aerial photograph of cataracts produced by the Bonneville Flood along the Snake River Valley reproduced with permission of Jim O'Connor, U.S. Geological Survey.

Figure 9.3   Graph showing the rise and fall of Lake Bonneville based on Godsey et al. 2005, fig. 3, p. 422.

Figure 9.4   Map of Lake Bonneville at its maximum size based on Godsey et al. 2005, fig. 2, p. 421.

Figure 10.1   Ediacaran-Cambrian time scale based on Marshall 2006, fig. 1.

Figure 10.2   Photographs of Ediacaran fossils from Wikimedia Commons, made available under the terms of the GNU Free Documentation License.

Figure 10.3   Images of Cambrian fossils reproduced courtesy of the Smithsonian Institution. Photographs of *Marrella, Hallucigenia,* and *Opabinia* from Briggs, Erwin, and Collier 1994. Ink drawings of *Hallucigenia* and *Opabinia* by Mary Parrish, Smithsonian Institution; ink drawing of *Marrella* by Laura Fry, Smithsonian Institution.

Figure 10.4   Times of animal phyla origination adopted from Prothero, 2007, figure 5.7, p. 138; reproduced with permission of Columbia University Press, based originally on fig. 2 in Briggs and Fortey 2005, p. 102.

Figure 11.1   Photograph of Devils Gate, Wyoming, reproduced with permission of Will K. Reeves.

Figure 11.2   Aerial photograph of Bighorn River through Sheep Mountain, Wyoming, reproduced with permission of Louis Maher.

Figure 11.5 (top) Virtual terrane image looking obliquely southwest over Split Mountain, Utah, from Google Earth. Image USDA Farm Service Agency, © 2010 Google, Image © 2010 DigitalGlobe, Image State of Utah.

Figure 11.5 (bottom) Photograph of the entrance to Split Mountain, Utah, from river level reproduced with permission of Steve Donaldson.

Figure 11.6 (bottom) Aerial photograph of the sub-summit surface on the southwestern face of the Wind River Range, Wyoming, reproduced with permission of David Lageson.

Figure 11.7   Geologic time panels illustrating the superimposition history of Rocky Mountain rivers modified from Mears 1993, figs. 2, 3 4, 6, 7, pp. 613–614; reproduced with permission of the Wyoming State Geological Survey.

Figure 12.2   Map of the Foreland Ranges modified from Meldahl 2007, fig. 7.9, p. 131; reproduced with permission of the University of Chicago Press.

Figure 12.3   Wind River Range and Medicine Bow Range cross-section diagrams based on Mears et al. 1986, fig. 3, p. 7, and fig. 20, p. 33.

Figure 12.4:   Diagrams of Wyoming before and after the Laramide Orogeny modified from Blackstone 1988, fig. 34, p. 49; reproduced with permission of the Wyoming State Geological Survey.

Figure 12.5 (bottom): Virtual satellite image of the Pampean Ranges of Argentina reproduced with permission of William Bowen, California Geographical Survey.

Figure 13.2   Gangplank drawing modified from Blackstone 1988, fig. 67, p. 117; reproduced with permission of the Wyoming State Geological Survey.

Figure 13.4   Map of mantle temperatures under the western United States based on Burdick et al. 2008, fig. 5f, p. 388.

Figure 13.5   Plate tectonics time diagrams explaining the Exhumation of the Rocky Mountains modified from Meldahl 2007, fig. 10.8, p. 200; reproduced with permission of the University of Chicago Press.

Figure 13.6   Cross-section view of the mantle and the Farallon Plate under the United States based on Sigloch, McQuarrie, and Nolet 2008, fig. 1, p. 459, and created in consultation with Karin Sigloch.

Figure A.3   Grand Canyon time scale based on Marshak 2008, fig. 12.10, p. 428.

Figure A.5   Siccar Point photograph by Dave Souza from Wikimedia Commons, made available under the terms of the GNU Free Documentation License.

INDEX

Abbey, Edward, xvi

Absaroka Mountains, 197n2

acanthite, 119

accreted terranes, 9 (fig.), 23–28, 27 (fig.),
    34n2, 48, 49, 52–53, 90, 92 (fig.),
    121, 124 (fig.), 147, 187

accretionary wedge, 7 (fig.), 10–14, 27 (fig.), 28,
    51 (fig.). *See also* Franciscan Complex;
    subduction

age of the Earth, 225–228

age of Earth's oldest rocks, 188, 228

Alps, 17, 18n4, 94 (fig.), 175, 207, 226

Altai Floods (Russia), 136

American River, 20, 36, 39, 49

Andes Mountains, 91, 184–185, 191–192

antecedent theory of rivers, 174, 176–177, 179, 182

Appalachian Mountains, 92, 104, 175, 213 (fig.)

Arkansas River, 80 (fig.), 168, 169 (fig.), 201
    (fig.), 202, 206

arthropods, 153–154, 156

Auburn, accreted terranes near, 235

auriferous gravels, 38–44, 49, 87, 91, 92 (fig.),
    124 (fig.)
    GPS locations for, 235, 236

Banks, John Edwin, 166

basalt. *See* Bonneville Flood; ophiolites; pillow
    basalt; Snake River Plain

basement-cored uplifts, 187–189, 191–192

basement rock, 187–189, 205
    GPS locations for, 242–243

Basin and Range Province, xv–xvi, 66 (fig.),
    71, 72 (fig.), 73, 75, 84–87, 90–93,
    95 (fig.), 98–114, 208, 209 (fig.)
    aerial view of, 82–83
    description of, 84–85
    faults of, 71, 73, 85 (fig.), 104–112
    formation of, 86–87, 90–93, 95 (fig.)
    magnitude of earthquakes in, 110–111
       (tables)
    mantle temperature beneath, 208, 209 (fig.)
    mountain uplift during earthquakes in, 87,
       98–114, 114n3
    range-roving rivers of, 179–180
    relationship to the Great Basin, 84–85, 93
    seismic belts of, 101–103
    seismic risk in, 105–112
    stretching of, 71, 72 (fig.), 73, 84, 101–103,
       185–186

Bates Mountain Tuff, 128
    GPS location for, 239
batholith. *See* Sierra Nevada Batholith
Battery, 129, Marin Headlands, 13–14, 15 (fig.)
    GPS location for, 233
Bear River, 36, 39, 49
Beartooth Mountains, 184, 186 (fig.)
Bighorn Range, 167, 184, 186 (fig.)
Bighorn River, 167, 168 (fig.)
bilateran animals, evolution of, 150, 157–161.
        See also Cambrian Explosion
Bingham Canyon Mine, 128
    GPS location for, 240
biomineralization, evolution of, 152
Black Hills, 184, 186 (fig.)
Black Rock Desert, 140
Bonneville Flood, 132–137. *See also* Lake
        Bonneville
    description of, 134
    Red Rock Pass outlet of, 133–134, 137, 138 (fig.)
    size compared to other floods, 134
Bonneville Speedway, 140–141
brachiopods, 155
braided streams, 208, 215n5
Breen, John, 145
Brown, James Berry, 163
Buenaventura River, 80–90, 94–96. *See also*
        Frémont, John C.
Buffon, Comte de (Georges-Louis Leclerc), 225
buoyancy of subducting plates. *See* subduction
burrowing, evolution of, 150 (fig.), 152, 159

calderas in the Great Basin, 123–128, 193
California Trail, 145
California, why it won't sink into the ocean,
        76n5
Cambrian Explosion, 148–161
    Darwin's thoughts about, 148–149
    evolution of chordates and, 160–161
    in the House Range, 153–156
    importance for animal evolution, 149, 157
    origin of animal phyla during, 156–157,
        158 (fig.)
    possible causes, 157–160
    preludes to, 149–152
Cambrian Period, 147–148
Camptonville, 44
Canadian Shield, 188

Cape Mendocino, 66 (fig.), 211
carbon-14 (radioactive carbon) in geologic
        dating, 67, 107, 111, 136, 145n1, 229 (fig.)
Carlin Canyon, 179, 218, 219 (fig.)
    unconformity in, 218–221, 231n3
Carlin Trend, 128
Carrizo Plain, 62 (fig.), 67, 68 (fig.)
Carroll, Sean, 161n4
Carson, Kit, 86, 89
Carson Pass, 57, 89, 94
    GPS location for, 237
Carson Range, 87–88, 105–110
Carson River, 86, 96
Cascade Range, 86, 95 (fig.), 100
Caucasus Mountains, 186
Central Nevada seismic belt, 101, 102 (fig.), 104
Central Pacific Railroad. *See* transcontinental
        railroad
Central Valley (of California). *See* Great Central
        Valley
Chengjiang Fauna (China), 153, 158 (fig.), 160
chert, 10, 11 (fig.), 13, 14, 15 (fig.), 17, 23
Cheyenne Tableland, 199–204, 208–210. *See
        also* Gangplank
Chocolate Mountains, 69
chordates, evolution of, 156, 158 (fig.), 160–161.
        See also Cambrian Explosion
cirques, 54, 55
climate change as a cause of the Exhumation,
        206–207, 210
Coast Range Ophiolite, 23, 27 (fig.), 28
Coast Ranges, 18n4, 23, 25 (fig.), 28, 35, 45n2,
        47, 51 (fig.)
Cocos Plate, 95 (fig.), 211
coevolution hypothesis for the Cambrian
        Explosion, 159–160
Colorado, elevation of, 183–184
Colorado Piedmont, 200 (fig.), 204 (fig.),
        205–206, 214n3
Colorado Plateau, xv, 83 (fig.), 84, 95 (fig.), 110,
        168, 170, 176–177, 179, 186 (fig.), 193
Colorado River, xv, 84, 168, 170, 176, 177, 179,
        184, 206, 217
Columbia River, 79, 86, 134, 184
Comstock Lode, 117–120, 128
    GPS location for, 239
contraction theory of mountains, 174–176
convection inside the Earth, 194, 226–227

Conway Morris, Simon, 153, 155
Cooke, Lucy, 166
Copernicus, 230
Cordillera. *See* North American Cordillera
Cornell, William, 165
Curie, Marie, 227
cyanobacteria, 150

Dakota Hogback, 205
Darwin, Charles, 148–149, 159, 160–161, 230
Dead Sea, 138, 146n2
Death Valley, 84, 148
deep time. *See* geologic time
Delano, Alonzo, 20, 21, 29, 30 (fig.), 166
density of oceanic plates, 194–195
Devils Gate, 166–167
    GPS location for, 241
Dickinson, William, 28
Dixie Valley fault, 104–106
    GPS location for, 238
Dodge, General Grenville, 202, 214n1
Donner, Eliza, 142–145
Donner Lake, 55
Donner Party
    crossing the Salt Lake Desert, 141–145
    in the Sierra Nevada, 55–57
Donner Pass, 55–59
    GPS location for, 236
Durant, Will, 35

Earth, age of, 225–228
earthquakes
    1857 Fort Tejon, 65, 66 (fig.), 100
    1872 Lone Pine (Owens Valley), 99–101
    1906 San Francisco, 15, 65, 66 (fig.), 100
    1915 Pleasant Valley, 104, 105 (fig.)
    1954 Dixie Valley, 104–105, 106 (fig.)
    1989 Loma Prieta, 65
    along the San Andreas fault, 63–71
    evaluating risk from, 105–112
    imaging the Earth's interior using, 191,
        207–208, 209 (fig.), 211, 213 (fig.)
    magnitude in the Basin and Range,
        110–111 (tables)
    mountain uplift during, 100–114, 114n3
    relation to tectonic plates, 63–65
Eastern California Shear Zone, 66 (fig.), 70
    (fig.), 71, 73 (fig.), 102 (fig.)

Eastern Sierra fault system, 42 (fig.), 87–88,
    114n2
East Pacific Rise, 5 (fig.), 7, 18n1
Ediacara Hills (Australia), 149
Ediacaran Fauna, 149–152, 157, 159
Eisenhower, Dwight, 202
elevations, ancient
    from fossil plants, 41–43, 206
    from fossil raindrops, 43
    from tilted sedimentary layers, 208–209
elevations of U.S. states, 183–184, 197n1
Empire Mine Historic Park, 234
erosion
    effect of climate on, 210
    effect over geologic time of, 103–104, 217
    exposure of granite by, 54
    exposure of metal ores by, 121
    unconformities caused by, 219–224, 226
    uplift caused by, 44, 207
Eurasian Plate, 61–63
Exhumation of the Rocky Mountains,
        170, 176–177, 180–181, 195–197,
        199–214
    effect on western Great Plains, 199–206
    Farallon Plate and, 210–214
    possible causes of, 206–214
    relationship to range-roving rivers, 176–177,
        180–181
    timing of, 215n4

Fairview Peak fault, 104–105
    GPS location for, 238
Farallon Islands, 3–5
Farallon-Pacific Ridge, 93, 95 (fig.), 210–211,
    212 (fig.)
Farallon Plate, 10–14, 17–18, 27 (fig.), 28–29,
        50–51, 90–95, 121, 123–124, 184,
        190–195, 210–214
    assembly of California by, 28
    caldera eruptions in the Great Basin and,
        123–128, 193
    exhumation of the Rocky Mountains and,
        210–214
    formation of the Basin and Range and,
        90–95
    origin of the Franciscan Complex and, 10,
        11 (fig.), 14, 17
    origin of metal ores and, 121–128

Farallon Plate (*continued*)
    origin of the Sierra Nevada Batholith and, 50–51
    under North America today, 211, 213 (fig.), 214
    uplift of the Rocky Mountains and, 91, 93 (fig.), 124 (fig.), 184, 190–195, 209 (fig.)
fault scarps, 98 (fig.), 99, 101, 104, 105 (fig.), 106 (fig.), 112 (fig.)
faults, measuring past earthquakes on, 67–70, 105–112
Feather River, 20, 36, 39, 96n3
Feather River Belt (ophiolite), 23
    GPS locations for, 234–235
Flaming Gorge, 171
fold-and-thrust belt, 91, 92 (fig.)
Foreland Ranges. *See* Rocky Mountains
fossil riverbeds. *See* auriferous gravels
fossils
    Cambrian of the House Range, 147–148, 153–156
    Pleistocene of the Great Basin, 133
    trace fossils, 152
Franciscan Complex, 10, 11 (fig.), 12 (fig.), 23, 27 (fig.), 45, 90. *See also* accretionary wedge
Frémont, John C.
    crosses the Sierra Nevada in winter, 89–90
    discovers the Great Basin, 89–90, 96
    exposes myth of Buenaventura River, 94–96
    report about Salt Lake Valley by, 114
    search for Buenaventura River, 85–90
Freud, Sigmund, 230
Front Range, xv, 168, 184, 186 (fig.), 188, 191, 201 (fig.), 203, 204 (fig.), 205–206
future geography of the western U.S., 74–75

gabbro, 11 (fig.), 22–23, 187, 194
Ganges River, 179
Gangplank, 200 (fig.), 203–204, 208, 214. *See also* Cheyenne Tableland
    GPS location for, 243
Genoa fault, 87–88, 90, 105–110, 111 (table)
    GPS location for, 237
    seismic risk from, 107–110

genome of bilateran animals, 157–158, 160, 161n4. *See also* Cambrian Explosion
geologic time
    compared to human time, xvii, 4, 8, 55, 103–104, 217–218, 221, 228–230
    "deep time" phrase origin, 230n1
    developments in the West and, xii–xiii
    discovery of, 217–231
    erosion's effects over, 103–104, 217
    James Hutton and, 221–224
    measurement using radioactivity, 227–228, 229 (fig.), 231n8
    plate movement over, 217–218
    rates of seafloor spreading and, 18n2
    unconformities and, 219–224, 226
    William Thomson and, 225–227
geosynclinal theory of mountains, 175
geothermal gradient, 226
glaciations
    Lake Bonneville and, 133, 136–137
    prehistoric floods and, 136
    in the Rocky Mountains, 178 (fig.), 210
    in the Sierra Nevada, 54–55
Global Positioning System (GPS) used to measure earth movements, xvii, 63, 72 (fig.), 73, 76, 101, 102 (fig.), 207
Golconda Terrane, 187
gold
    amount mined in California, 39 (fig.)
    amount mined in Nevada, 117
    auriferous gravels, 38–44, 49, 87
    concentration in Earth's crust of, 21
    hydraulic mining of, 36–40
    lode gold formation, 28–30
    lode gold mining, 30–34
    lode versus placer defined, 21
    placer gold formation, 38
    placer gold mining, 36–40
Golden Gate, 4, 10, 14, 40, 210
    GPS locations for, 233–234
Golden Gate Bridge, 13–16
Gore Range, 168
Gould, Stephen Jay, 153, 217, 230
Grand Canyon, xvii, 168, 170, 177, 179, 188, 206, 221, 222 (fig.), 223 (fig.)
    erosion of, 206, 217, 231n2

John Wesley Powell's journey through, 177–179
  unconformities in, 221–223
Grand Wash Cliffs, 179
granite
  beneath the Andes, 54
  erosion of, 54–57
  on the Farallon Islands, 3–4
  origin during subduction, 50–52
  in the Rocky Mountains, 165–167, 187–188, 203 (fig.)
  in the Sierra Nevada, 11 (fig.), 27 (fig.), 28, 49–59, 59n1, 87, 92 (fig.)
Grass Valley, 23
Grayback Hills, 142, 143 (fig.)
Great Basin, 40–43, 55, 57, 82–87, 90–93, 117–123
  aerial view of, 82–83
  description of, 40, 84–85
  discovery of, 89–90, 96
  earthquakes in, 104–112
  faults of, 85 (fig.), 88 (fig.), 104–112
  geologic formation of, 40–43, 85 (fig.), 86–87, 88 (fig.), 90–93
  metal ores in, 117–123
  Pleistocene lakes in, 136–137
  relationship to the Basin and Range, 84–85
  volcanism in, 90–91, 123–128, 193
Great Central Valley, 28, 35, 40, 44, 47, 71, 96
Great Plains, xv, xvi, 91, 165, 167–168, 184, 195, 197, 199–206, 210
Great Salt Lake, 84, 114, 132, 137–139
Great Unconformity (Grand Canyon), 221, 223 (fig.)
Great Valley Group, 27 (fig.), 28
Great Valley Ophiolite, 23, 27 (fig.), 28, 92 (fig.)
Green River, 170–174, 184, 188, 189
groundwater and metal ore formation, 29–30, 119–123
Gulf of California, 73 (fig.), 75, 76n2, 95 (fig.), 211

Hastings Cutoff, 141, 145
Hastings, Lansford, 141–143, 145
High Plains, 200 (fig.), 204–205
High Plains Escarpment, 201 (fig.), 204 (fig.), 205
  GPS location for, 243

Himalaya, 18, 100, 179
Hope Wells, 141, 145
House Range, 147–148, 153
  GPS locations for, 240
Humboldt River, 84, 145, 179, 182, 218, 219 (fig.)
Hutton, James, 50, 221–224, 225, 226, 231n4
hydraulic mining, 36–40
hydrothermal activity
  Great Basin ore formation and, 119–123
  in Iceland, 61
  lode gold and, 29–30, 31 (fig.)
  mercury deposits and, 45n2
  at mid-ocean ridges, 34n3
  serpentinite and, 16 (fig.), 17

ice ages. See glaciations
iceberg analogy for mountains, 207
Iceland, 61–63, 90
Indians of the Great Plains, 202
Intermountain seismic belt, 101, 102 (fig.)
Interstate highway system, 202
isostatic equilibrium, 76

Jack Hills (Australia), 228
Juan de Fuca Plate, 12 (fig.), 73 (fig.), 95 (fig.), 100, 209 (fig.), 211
Juan Fernández Rise, 192 (fig.), 195

Kelvin, Lord. See Thomson, William (Lord Kelvin)
Koppeh Dagh Mountains, 186
Krakatoa, eruption of, 128

Lake Bonneville, 132–137, 138 (fig.)
  adjustment of crust under, 137
  Bonneville Flood and, 132–136
  Pleistocene fauna around, 133
  rise and fall of, 136 (fig.)
  shorelines of, 136–137
Lake Combie Ophiolite, 48
  GPS location for, 235
Lake Lahontan, 140
Lake Tahoe, 88 (fig.), 107, 110
Laramide Orogeny, 91, 177, 184, 189–190, 191–195, 213 (fig.), 214. See also Rocky Mountains

Laramie Range, 165, 168, 184, 186 (fig.), 188,
    189 (fig.), 199–204
    GPS location for, 241–242
Laramie River, 167–168
Lawson, Andrew, 15, 17
Lewis and Clark expedition, xviii, 79
Linnaeus, Carolus, 161n2
lithosphere, 75n1
lode gold. *See* gold
Lodgepole Creek, 200 (fig.), 202
Lodore Canyon, 171, 172 (fig.)
    GPS location for, 241
Lone Pine fault, 97–100, 101, 114n2
    GPS location for, 238
Lone Pine (Owens Valley) earthquake of 1872,
    99–101, 110 (table)
Longs Peak, 188
Los Trancos Preserve (San Andreas fault site),
    69
    GPS location for, 237
Louisiana Purchase, 79

magma
    chemical differentiation of, 50–51, 59n2
    effect of cooling rate on crystal size in, 52
    ore formation and, 121–123
    origin of during subduction, 50–51
    role of water in forming, 50, 51 (fig.), 193
Malakoff Diggins, 36–37
    GPS location for, 235
mantle, xvii, 7 (fig.), 11 (fig.), 17, 18n3, 23
    convection in, 226–227
    imaging of. *See* seismic tomography
    magma formation in, 50, 51 (fig.), 59n2
    physical nature of, 197n3
    plumes in, 63
    temperature under western U.S. of,
    207–208, 209 (fig.)
mantle drip beneath the Sierra Nevada, 209 (fig.)
Marin Headlands, 6 (fig.), 10, 13, 15 (fig.)
Marvine, Archibald, 176–177
Massacre Rocks, 131–132. *See also* Bonneville
    Flood
    GPS location for, 240
McCall, A.J., 55, 166
McMillan, Margaret, 208, 210
McPhee, John, 1, 49, 77, 182n2, 197n3, 221,
    230n1

Medicine Bow Mountains, 184, 186 (fig.)
    GPS location to view sub-summit surface
    of, 241–242
Mediterranean Sea, filling of, 134
melon gravels, 132, 134. *See also* Bonneville
    Flood
mercury in gold mining, 33, 38, 39, 45n2
meteorites and age of the Earth, 228, 230 (fig.)
Mid-Atlantic Ridge, 5 (fig.), 7, 61–63, 90
Middleton, Joseph, 166
mid-ocean ridges. *See* seafloor spreading
Mississippi River, 129n1, 206
Missoula Floods, 136
Missouri River, 59, 79, 184
mobility in animal evolution, 152. *See also*
    Cambrian Explosion
molecular clock, 161n1
Moore, Martha, 166
Moores, Eldridge, 17, 28
Mormons, 112–114, 179
Mother Lode gold district, 29–34, 48, 123
mountain belts, location of, 184–187
mountain uplift, 91–92, 94 (fig.), 100, 103
    earthquakes and, 100–114
    slow- versus fast-rising, 107, 108 (fig.)
Mount Evans, 201 (fig.), 205–206
    GPS location for, 243
Mount Whitney, 97–100, 114n1
Muir, John, xv, xvi, 47–48, 54, 77, 100

Nazca Plate, 184–185, 191–192, 195
Neenach volcanics, 69, 70 (fig.)
Nevada, metal wealth of, 117
Nevadaplano, 40–43, 49, 86–87, 90–91, 92
    (fig.), 114n3, 124 (fig.)
North America, geologic assembly of,
    187–188
North American Cordillera, xvii, 90, (92 (fig.),
    184, 185 (fig.)
North American Plate, 11 (fig.), 12 (fig.), 61–76,
    95 (fig.), 102 (fig.), 210, 212 (fig.),
    213 (fig.)
    eastern edge of, 61–62
    movement of relative to the Pacific Plate,
    63–75
    western edge of, 73–75
North Platte River, 168, 184, 200 (fig.), 202,
    206

North Star Mining Museum, 234
Nuvvuagittuq Belt, 188, 228

oceanic plateaus as a cause of flat subduction,
    91, 195
oceanic plates, density and buoyancy of,
    194–195
ocean trenches. *See* subduction
Ogallala Formation, 201 (fig.), 203, 205, 208
Omega Diggins, 36
    GPS location for, 235
Ontong-Java Plateau, 195
onychophorans, 155–156
ooze, 7–8, 10
ophiolites, 9 (fig.), 11 (fig.), 17, 18n4
    in the Alps, 17
    in the Coast Ranges, 23, 27 (fig.), 28
    at the Golden Gate, 11 (fig.), 17
    and lode gold formation, 23, 28–30
    in the Sierra Nevada, 22–23, 24 (fig.),
        25 (fig.), 27 (fig.), 48
    See also Coast Range Ophiolite; Feather
        River Belt (ophiolite); Franciscan
        Complex; Lake Combie Ophiolite;
        Smartville Ophiolite
ophiolitic suite, 11 (fig.), 17
ore formation
    association with magma and faults, 119–123
    in the Great Basin, 117–123
orogeny. *See* Basin and Range Province;
        Laramide Orogeny; mountain uplift;
        Sevier Orogeny
Owens Valley (Lone Pine) fault, 97–100
    GPS location for, 238
Owl Creek Mountains, 167, 184, 186 (fig.)
oxygen and the Cambrian Explosion, 157, 161n3

Pacific Plate
    formation of the San Andreas fault by,
        93 (fig.), 95 (fig.)
    movement relative to North America,
        65–66, 75–76
    relationship to the North American Plate,
        63–75
    stretching of the Basin and Range by,
        101–103
Pampean Ranges (Argentina), 191–193
Pangaea, xvii, 8, 29, 90, 104

Park Range, 168, 184, 186 (fig.)
passive uplift caused by erosion, 207
pediments, 177, 181 (fig.), 196 (fig.), 214n2.
    See also sub-summit surfaces
peridotite, 5 (fig.), 11 (fig.), 16 (fig.), 17, 23, 25
    (fig.), 194. *See also* serpentinite
Perry, John, 226–227
Peru-Chile Trench, 5 (fig.), 51 (fig.), 185, 191, 192
    (fig.), 195
phyla, origin of animal, 156–157, 158 (fig.)
pies sold to gold rush miners, 20
pillow basalt. *See also* seafloor spreading
    formation of, 5–8
    at the Golden Gate, 5, 6 (fig.), 8, 10, 14, 17
    in the western Sierra Nevada, 21–22
Pilot Peak (Pilot Range), 142, 143 (fig.), 144
    (fig.), 145
Pilot Spring (Donner Spring), 143–145
Pinnacles volcanics (Pinnacles National
    Monument), 69, 70 (fig.)
    GPS location for, 237
placer gold. *See* gold
plankton, 3, 7–8, 10, 13, 14
plates, 4, 5 (fig.), 7 (fig.), 11 (fig.), 61–64, 73
    (fig.). *See also* seafloor spreading;
        subduction
    cause of movement of, 194
    density and buoyancy of oceanic, 194–195
    earthquakes and, 63–66
    mountain uplift and, 91–92, 94 (fig.), 100,
        103, 174
    movement of, 4, 5 (fig.), 7 (fig.), 73 (fig.)
    world map of, x–xi
plate tectonics, theory of, 4, 50, 174, 175, 182n2
Platte River, 215n5
Playfair, John, 222, 224
plutons, 34n2, 51 (fig.), 52, 91, 92 (fig.)
    ore formation around, 121–123, 128
Point Bonita, 4, 5, 8, 10
    GPS location, 233
polychaete worms, 155
Pompeii, 127 (fig.), 128
Pope, General John, 202
Portneuf River, 133
Powell, John Wesley, 163, 170–182, 189
    antecedent theory of, 174, 176–177, 179, 182
    contraction theory and, 174–176
    early life of, 170

Powell, John Wesley (*continued*)
    expedition down Green and Colorado
        rivers, 170–174, 177–179
    puzzlement over range-roving rivers,
        171–172
priapulid worms, 155
Promontory Summit, 59
Pyramid Lake, 84
pyroclastic flows, 125–128

quartz veins. *See* veins

radioactivity
    age-dating rocks using, 227–228, 229 (fig.),
        231n8
    age of the Earth based on, 228
    discovery of, 227
    heat inside the Earth and, 175, 182n1
radiolarians, 13, 14
rapids, cause of, 171
Rattlesnake Creek Pluton, 52, 53 (fig.)
Red Rock Pass, 133–134, 137, 138 (fig.)
    GPS location for, 240
Register Rock, 131
    GPS location for, 240
Rio Grande Rift, 83 (fig.), 95 (fig.)
Rocky Mountains, xvi, xviii, 79, 80–84, 91,
        92 (fig.), 110, 123, 124 (fig.), 163–215
    basement rock of, 187–189
    Buenaventura River and, 80–84
    burial of, 181 (fig.), 196–197
    early exploration of, 79–84
    exhumation of, 170, 176–177, 180–181,
        195–197, 199–214
    Foreland Ranges within, 184, 185 (fig.),
        186 (fig.), 190
    GPS locations for basement rock outcrops
        in, 242–243
    Pampean Ranges as analog for, 191–193
    pioneer impressions of, 165–166
    puzzle of location, 186–187
    puzzle of rivers that flow through, 166–177
    sub-summit surfaces within, 177, 178 (fig.),
        181 (fig.), 189 (fig.), 196 (fig.), 204 (fig.),
        205–206
    uplift of during Laramide Orogeny, 91, 177,
        184, 189–190, 191–195, 213 (fig.), 214
    western Great Plains and, 199–206

Roller Pass, 57
Ruskin, John, 163, 217

Sacramento River, 39, 78
Salinian Block, 4
Salt Lake City, 111–114
    seismic risk from Wasatch fault, 111–114
    settlement of, 113–114
Salt Lake Desert, 139–145
    Bonneville Speedway on, 140–141
    Donner Party crossing of, 141–145
    formation of, 139
    Mark Twain's description of, 139–140
San Andreas fault, 4, 14, 62 (fig.), 63, 65–76,
        93 (fig.), 93, 101, 102 (fig.), 124 (fig.), 185,
        210, 212 (fig.)
    formation of, 93, 95 (fig.)
    by the Golden Gate, 14–15
    at Los Trancos Preserve, 69
    movement of Farallon Islands by, 4
    at Pinnacles National Monument, 69, 70
        (fig.)
    at Wallace Creek (Carrizo Plain), 67–69
Sangre de Cristo Range, 184, 186 (fig.), 201 (fig.)
San Joaquin River, 40, 81
San Juan Mountains, 197n2
Sarychev Volcano, 126 (fig.)
Sawatch Range, 184, 186 (fig.)
Sawyer, Judge Lorenzo, 40
scarps. *See* fault scarps
Scout, very good dog, 32 (fig.)
seafloor spreading, 5 (fig.), 7 (fig.), 7–8, 18n2,
        93, 193, 211
Sedgwick, Adam, 148
seismic belts, 101, 102 (fig.)
seismic reflection, 106 (fig.)
seismic risk, evaluation of, 105–112
seismic tomography, 191, 207–208, 209 (fig.),
        211, 213 (fig.), 214
seismic waves. *See* earthquakes
seismometers, 207–208
serpentinite, 11 (fig.), 14–17, 45n2
    as California state rock, 18
    at the Golden Gate, 14–18
    GPS location for Golden Gate outcrops of,
        234
    origin of, 17
    in the Sierra Nevada, 23, 25 (fig.), 50, 187

Sevier Orogeny, 91

Shatsky Rise, 197n4

Sheep Mountain, 167, 168 (fig.)

Sherman Granite (Laramie Range), 188, 189
    (fig.), 203
    GPS location for, 242

Sherman Summit, 203

Sherman, William Tecumseh, 202

Siccar Point (Scotland), 224, 231n4

Sieh, Kerry, 67

Sierra Madre Range, 184, 186 (fig.)

Sierra Nevada, xviii, 3–4, 11 (fig.), 19–34, 35–44,
    47–59
    fossil riverbeds of. *See* auriferous gravels
    glaciations in. *See* glaciations
    gold in. *See* gold
    granite in. *See* granite
    John Muir's description of, 47–48
    relationship to Farallon Islands, 3–4
    relationship to the Nevadaplano, 40–42
    Sierran Plate and. *See* Sierran Plate
    terranes in. *See* accreted terranes
    uplift of, 42 (fig.), 43–44, 99–100

Sierra Nevada Batholith, 27 (fig.), 28, 49, 51
    (fig.), 52, 53 (fig.), 54, 91, 92 (fig.), 122
    formation of, 52–59, 91
    GPS location for western edge, 236
    western edge near Big Bend, 52–54

Sierran Plate, 62 (fig.), 68 (fig.), 70 (fig.), 71–73,
    102 (fig.), 114n2

Sigloch, Karin, 211

slab pull, 194

slab window, 211

sluices, 33, 38, 39

Smartville Ophiolite, 22–23
    GPS location for, 234

Smith Jedediah, 85

Snake River, 84, 131–134, 135 (fig.). *See also*
    Bonneville Flood

Snake River Plain, xvii, 130–134, 195

South Pass, 79, 81, 89

South Platte River, 168, 200 (fig.), 202, 206

Split Mountain, 171–173, 179
    GPS location for, 241

sponges, 155–157

Sprigg, Reginald, 149

stamp mill, 32 (fig.). *See also* gold: lode gold
    mining

Stanford, Leland, 21

Steinmann, Gustav, 17, 18n4

Steinmann Trinity, 11 (fig.), 17–18

Stephens, Elijah, 57

Stewart, Robert, 79

Stillwater Range, 104–106
    GPS location for, 238

subduction, 5 (fig.), 7 (fig.), 8, 10, 11 (fig.),
    12 (fig.), 13, 17–18
    accretionary wedges formed during, 10–14,
        11 (fig.), 12 (fig.)
    buoyancy of oceanic plates and, 194–195
    flat subduction of Farallon Plate, 91, 92
        (fig.), 191–195
    flat subduction under the Andes today,
        191–192
    magma formation during, 50–51, 193
    oceanic plateaus and, 195
    slab pull as driving force of, 194

sub-summit surfaces, 177, 178 (fig.), 181 (fig.),
    189 (fig.), 196 (fig.), 204 (fig.), 205–206

superimposition theory of rivers, 176–177, 181
    (fig.)

Sutter's Fort, 89, 94

Swain, William, 20

Sweetwater River, 166, 167 (fig.)

taxonomy, 156, 161n2

tectonic plates, world map of, x–xi

terranes. *See* accreted terranes

Teton Range, 197n2

thermal contraction theory of mountains,
    174–175

Thingvellir Rift Valley (Iceland), 62 (fig.)

Thomson, William (Lord Kelvin), 225–227,
    231n5, 231n6, 231n7

thrust faults, 51 (fig.), 180–181, 189, 190 (fig.),
    191

Tian Shan Mountains, 186

time. *See* geologic time

Tobin Range, 104, 105 (fig.)

Trans-Antarctic Mountains, 177

transcontinental railroad, xvi, 48, 57, 59, 199,
    202

Transverse Ranges, 66 (fig.), 69, 70, 73
    (caption), 82 (fig.)

trenches. *See* subduction

Truckee River, 86, 96

tuff (volcanic), 59n2, 70 (fig.), 125 (fig.), 128
Twain, Mark, 84, 118, 129n1, 129n3, 139–140

Uinta Mountains, 171–174, 184, 186 (fig.), 188
unconformities, 219–224, 226. *See also*
    geologic time
Union Pacific Railroad. *See* transcontinental
    railroad
Ural Mountains, 186
USArray seismometer grid, 207–208, 209
    (fig.), 211, 213 (fig.)
Ussher, Archbishop James, 217, 221, 228

veins
    lode gold in California within, 28–30, 34
    metal ores in the Great Basin within,
        119–123
Ventura Canyon, 179
Vesuvius, eruption of, 128
Virginia City, 117–119
Virginia Mountains, 118
volcanic rocks
    in the Great Basin, 91, 123–128
    on the ocean floor. *See* pillow basalt
    offset by the San Andreas fault, 69, 70
    in the Sierra Nevada, 49, 51

Walker, Joseph, 85
Walker Lake, 84, 96n1
Walker Lane fault system, 66 (fig.), 70 (fig.), 71,
    72 (fig.), 73, 75, 101, 102 (fig.), 105, 114n2

Walker River, 87, 96n1
Wallace Creek (San Andreas fault site),
    67–69
    GPS location for, 236
Warren, Robert Penn, xviii
Wasatch fault, 107, 109 (fig.), 110–114
    GPS locations for, 238–239
Wasatch Range, 84, 85 (fig.), 109 (fig.), 107–114,
    133, 137, 139, 141
Wegener, Alfred, vi, 175
Werner, Abraham, 50
Wet Mountains, 168
Wheeler, John, 218
Wheeler Shale, 148, 153, 158 (fig.)
Whitney, Josiah, 48
Wilson, William, 165
Wind River Canyon, 167
    GPS location for, 241
Wind River Range, 178 (fig.), 184, 186 (fig.),
    188, 189 (fig.)
    GPS location for basement rocks within,
    242
Wolcott, Lucien, 166
Wyoming, elevation of, 183–184

xenoliths, 52–53, 54 (fig.)

Yampa River, 171, 172 (fig.)
Yosemite Valley, 55, 100
Young, Brigham, 114, 114n4
Yuba River, 21, 36, 37, 39, 44

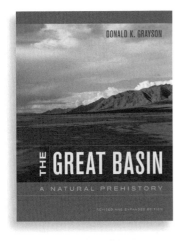

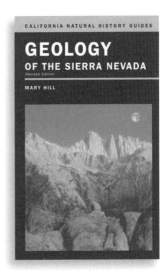

The Great Basin

A Natural Prehistory
DONALD K. GRAYSON
Revised and Expanded Edition

"A grand synthesis and it represents a major revision written in an engaging style that teaches on every level." **—Steven R. Simms, author of *Ancient Peoples of the Great Basin and the Colorado Plateau***
$75.00 cloth   978-0-520-26747-3

Geology of the Sierra Nevada

MARY HILL
Revised Edition

"Hill's masterful book...tells what the Sierra Nevada is made of, how it was and is being made, and how you might see the evidence of these things for yourself."**—American West**
California Natural History Guides
$21.95 paper   978-0-520-23696-7

A Land in Motion

California's San Andreas Fault
MICHAEL COLLIER

A Land in Motion provides a geologic tour of the San Andreas Fault in an accessible narrative punctuated with dramatic color illustrations, lively anecdotes, and authoritative information about earthquakes.
Copublished with Golden Gate National Parks Association
$25.95 paper   978-0-520-21897-0

Geology of the San Francisco Bay Region

DORIS SLOAN
Photography by John Karachewski

"Sloan plumbs the depths of the bay and scales surrounding hillsides, all the time explaining how this geological puzzle was formed over millions of years. She points out the easily overlooked everyday scenes that are significant if the viewer looks a little closer. After reading her guide, a trip to the Bay will never be the same." **—Sacramento Bee**
California Natural History Guides
$18.95 paper   978-0-520-24126-8

Why Geology Matters

Decoding the Past, Anticipating the Future
DOUG MACDOUGALL

"Macdougall does a masterful job of exploring the questions, dilemmas, and insights that have led to today's scientific understanding of the composition of our planet.... A compelling, interdisciplinary peek at Earth's prehistory—including those processes that support so much of modern civilization."
—Ernest Zebrowski, author of *Global Climate Change* and *Category 5: The Story of Camille*
$29.95 cloth   978-0-520-26642-1

COMPOSITION: Westchester
TEXT: 9.5/14 Scala
DISPLAY: Scala Sans
PRINTER AND BINDER: Thomson-Shore